Photopolymerization

Photopolymerization

Fundamentals and Applications

Alec B. Scranton, EDITOR
Michigan State University

Christopher N. Bowman, EDITOR
University of Colorado

Robert W. Peiffer, EDITOR
DuPont, Inc.

Developed from a symposium sponsored by the
Division of Polymeric Materials: Science and Engineering, Inc.
at the 211th National Meeting of the American Chemical Society,
New Orleans, Louisiana,
March 24–28, 1996

American Chemical Society, Washington, DC

Library of Congress Cataloging-in-Publication Data

Photopolymerization: fundamentals and application / Alec B. Scranton, editor, Christopher N. Bowman, editor, Robert W. Pheiffer, editor.

p.   cm.—(ACS symposium series, ISSN 0097–6156; 673)

"Developed from a symposium sponsored by the Division of polymeric Materials: Science and Engineering, Inc., at the 212th National Meeting of the American Chemical Society, New Orleans, Louisiana, March 24–28, 1996."

Includes bibliographical references and indexes.

ISBN 0–8412–3520–1

1. Photopolymerization—Congresses. 2. Photopolymerization—Industrial applications—Congresses.

I. Scranton, Alec B., 1963–   . II. Bowman, Christopher N., 1967–   . III. Peiffer, Robert W., 1942–   . IV. American Chemical Society. Division of Polymeric Materials: Science and Engineering. V. American Chemical Society. Meeting (211th: 1996: New Orleans, La.) VI. Series.

QD716.P5P48   1997
668.9′2—dc21   97–30337
                                                              CIP

This book is printed on acid-free, recycled paper.

PRINTED IN THE UNITED STATES OF AMERICA

Foreword

THE ACS SYMPOSIUM SERIES was first published in 1974 to provide a mechanism for publishing symposia quickly in book form. The purpose of the series is to publish timely, comprehensive books developed from ACS sponsored symposia based on current scientific research. Occasionally, books are developed from symposia sponsored by other organizations when the topic is of keen interest to the chemistry audience.

Before agreeing to publish a book, the proposed table of contents is reviewed for appropriate and comprehensive coverage and for interest to the audience. Some papers may be excluded in order to better focus the book; others may be added to provide comprehensiveness. When appropriate, overview or introductory chapters are added. Drafts of chapters are peer-reviewed prior to final acceptance or rejection, and manuscripts are prepared in camera-ready format.

As a rule, only original research papers and original review papers are included in the volumes. Verbatim reproductions of previously published papers are not accepted.

ACS BOOKS DEPARTMENT

Contents

Preface

PHOTOPOLYMERIZATION IS ONE SCIENTIFIC DOMAIN that offers both a wealth of tantalizing fundamental challenges that cut across many disciplines and compelling practical applications for successful investigators. The fundamentals and applications of photopolymerizations are exciting areas of active research. Over the past few decades, many challenges in the field of photopolymerization have been overcome, and the worldwide market for UV-curable systems approached $1 billion in 1995. However, the field is still relatively young and offers tremendous opportunities for creative scientists and engineers. For example, a fundamental characterization of the relationships among the kinetics, mechanisms, structure, and properties of existing photopolymerization systems will lead to optimization of a host of processes, and reaction systems that are only being contemplated today are destined to develop into major products in the future.

Photopolymerization is one of the most rapidly expanding processes for materials production with more than 15 percent annual growth projected for the next several years. Ultimately, the growth in applications of photopolymerization is driven by the advantages afforded by the use of light, rather than heat, to drive the conversion of monomer to polymer. These advantages include solvent-free formulations, very high reaction rates at room temperature, spatial control of the polymerization, low energy input, and chemical versatility because a wide variety of monomers can be polymerized photochemically. Because of this unique set of advantages, photopolymerizations have gained prominence in recent years for the pollution-free curing of polymer films as well as emerging applications in dental materials, conformal coatings, electronic and optical materials, and rapid prototyping of three-dimensional objects.

This volume is based on a four-session symposium presented at the 211th National Meeting of the American Chemical Society, sponsored by the ACS Division of Polymeric Materials: Science and Engineering, Inc., in New Orleans, Louisiana, March 24–28, 1996. The general interest in the topic of photopolymerization was illustrated by the excellent attendance and active discussions during the symposium, and we were delighted when most of the speakers agreed to contribute chapters to this book. The volume is well balanced, with contributions from both academia and industry, and should provide the reader with an excellent idea of the current directions of photopolymerization research.

The chapters in this book are organized into three sections. A majority of the commercial photopolymerization systems are based on acrylate monomers; therefore, the first several chapters focus on fundamental characterization of acrylate systems. One underlying theme of these chapters is the use of advanced analytical tools to characterize the reactions. Indeed, many of these tools could be applied to nonacrylate systems as well.

The second section focuses on emerging classes of photopolymerizations that are being developed as alternatives to acrylates. Three types of polymerization systems are included: cationic photopolymerizations, initiator-free charge-transfer polymerizations, and a thiol-ene reaction system. The last section covers four interesting emerging applications of photopolymerization technology.

Acknowledgments

We believe that the reader will find numerous examples of ingenious science and creative engineering in this volume, and we thank each of the authors for a job well done. In addition, we thank the editorial staff of the American Chemical Society as well as the secretarial staffs at Michigan State University, the University of Colorado, and DuPont for their support in this endeavor.

ALEC B. SCRANTON
Department of Chemical Engineering
Michigan State University
A202 Engineering Building
East Lansing, MI 48824–1226

CHRISTOPHER N. BOWMAN
Department of Chemical Engineering
University of Colorado
Campus Box 424
Boulder, CO 80309–0424

ROBERT W. PEIFFER
Experimental Station Laboratory
Dupont, Inc.
P.O. Box 80257, Route 141
Wilmington, DE 19880–0257

July 14, 1997

Chapter 1

Applications of Photopolymer Technology

Robert W. Peiffer

Experimental Station Laboratory, DuPont, Inc., P. O. Box 80257, Route 141,
Wilmington, DE 19880–0257

This review will highlight the interrelationships between basic
photopolymer science and practical applications of this technology.
Each application of photopolymer technology can be described in terms
of three primary descriptors; the mode of exposure, the mechanism of
the photopolymer reaction employed and the visualization method
used. Using this foundation, the widely diverse applications of
photopolymer technology to electronic materials, printing materials,
optical and electro-optical materials, the fabrication of devices and
polymeric materials, adhesives and coating materials will be discussed.

Photopolymer technology, which encompasses the action of light to form polymers
and light initiated reactions in polymeric materials, is an immense topic. Previous
papers in this symposium have described some of basic chemistry utilized in
photopolymer technology. The primary objectives of this paper are a) to develop the
connections between basic photopolymer chemistry and practical uses of the
technology and b) to provide an overview of the wide variety of photopolymer
applications that have been developed since the 1950's . Every attempt has been made
to make this review as inclusive as possible, but because of the extensive nature of this
topic, there are many applications of photopolymer chemistry that have not been
included. In addition, only limited representative references are provided since the
patent and open literature for this technology are quite vast (1).

Background

"Photopolymer Technology" encompasses those chemical and physical reactions of
organic-based materials that are initiated by the application of electromagnetic
radiation. Normally ultraviolet (UV), visible or infrared (IR) light are utilized to
initiate photopolymer reactions, but higher energy sources at shorter wavelength (e.g.
electron-beam) have also been utilized. The practical applications of photopolymer

chemistry have rapidly emerged during the last three decades.   One of the first demonstrations of photopolymer chemistry occurred early in the last century when metal plates were coated with a natural resin, exposed to sunlight and washed with a solvent to obtain crude images (2).   The practical development of organic-based imaging systems did not occur until the 1930's and 1940's when the foundations of modern photopolymer chemistry were developed.   The first synthetic photopolymer materials, poly(vinyl cinnamate) light sensitive polymers, were developed at Kodak (3).   A more complete demonstration of the versatility of polymeric imaging was achieved at DuPont with the development of photoinitiated free-radical chain polymerization (4).   Since the development of this basic chemistry, photopolymer technology has been utilized in a wide variety of practical applications which are the subject of this paper.   It has been estimated that the yearly worldwide market for photopolymer materials is in excess of 10 billion dollars (5).

Overview of Photopolymer Technology

The relationship between basic photopolymer chemistry and practical applications of this technology can best be described by the *exposure mode* of the given application (imaging and non-imaging), the *mechanism* of the photopolymer reaction employed and the *visualization method* utilized to detect changes brought about in the organic material through the action of electromagnetic radiation. Thus, a given application of photopolymer technology can be characterized by three basic descriptors:

> ***Exposure Mode***
> - Imaging
> - Non-Imaging
>
> ***Mechanism of Photopolymer Reaction Used***
> - Photopolymerization
> - Photocrosslinking
> - Photomolecular Reaction
> - Photodegradation
> - Photo/Thermal Reaction
>
> ***Visualization Method Used***
> - Solubility
> - Adhesion
> - Color Change
> - Phase Change
> - Refractive Index
> - Electrical Conductivity

The varied applications of photopolymer technology can be divided into two general classifications based on the *exposure mode* of the applied radiation.   The two exposure modes are imaging, where the photopolymer material is exposed in a "pattern-wise" process, and non-imaging, where the photopolymer is given an "overall" or non-pattern exposure.   Many of the high "value-in-use" applications are in an imaging category (e.g. film and liquid resists, printing plates) while the largest worldwide applications for photopolymer-based products are in a non-imaging classification (e.g. photosensitive protective and decorative coatings, adhesives, sealants).

All the applications of photopolymer technology rely on the formation of detectable changes in organic materials following exposure to light or other types of electromagnetic energy. The light initiates changes in the organic material by one of five mechanisms and the resultant chemical and/or physical change in the material is "readout" by one of the six primary visualization methods. As a result of these two steps, photopolymer materials can be designed to be quite sensitive detectors of small chemical and/or physical changes in solubility, adhesion balance, phase change, refractive index, conductivity, etc.

Basic Mechanisms of Photopolymer Reactions

The basic light initiated reactions used in the diverse applications of photopolymer technology can be classified into 5 categories based on the chemical and physical processes utilized.

Photopolymerization. The photoinitiated polymerization of monomers and oligomers to form crosslinked higher molecular weight material is the basis for most of the commercial applications of photopolymer technology. The reactive monomeric materials that are most commonly used are low molecular weight unsaturated acrylate or methacrylate monomers that can be made to crosslink with the use of a radical-generating photoinitiator (*6*). Cationic initiated crosslinking of monomeric materials with epoxy and/or vinyl ether functionality has increased in practicality with the development of new higher efficiency photoinitiators that generate cationic species (e.g. strong acids) upon exposure (*7*). Photopolymerization chemistries other than acrylate and epoxy/vinyl ether systems (e.g. polyesters and maleate/vinyl ether formulations) have also been used (*8-9*). Typical photopolymer formulations contain a photoinitiator system, monomers and oligomers, a polymer or polymers to provide specific physical and/or processing properties and a variety of additives to modify the physical properties of the light sensitive composition or the final properties of the cured photopolymer. Photopolymerization chemistry is commercially utilized in a wide variety of applications (e.g. coatings, adhesives, sealants, printed wiring board resists, microelectronic resists and printing plates).

Photocrosslinking. The second class of photopolymer chemistry that is used in some commercial products is based on the reaction of unsaturated moieties attached to an organic polymer. These photopolymer materials include the [2+2] cycloaddition of the ethylenic groups in poly(vinyl cinnamate) polymers and in the newer styryl pyridinium (*10*) and thiazolium (*11*) derivatives of poly(vinyl alcohol). The main advantage of this chemistry is that, unlike free-radical photopolymerization, they are insensitive to the presence of oxygen. This photopolymer mechanism is principally used in applications employing a washout development process (e.g. resists).

Photomolecular Reactions. The third basic mechanism of photopolymer chemistry relies on the light initiated reaction of small organic molecules to modify the polymer-based matrix into which they are incorporated. This class of photopolymer reactions is illustrated by the rearrangement of o-quinonediazide derivatives (*12*) in microresist materials and by the isomerization of azobenzene derivatives in optical storage devices (*13*). Other small organic materials have been used to modify the plasticization of a polymer medium to modulate tack (*14*) and/or adhesion and to initiate color changes following exposure to light (*15*). There are many examples of this class of photopolymer materials.

Photodegradation. The next general class of photopolymer reactions that can be used in a variety of applications is based on the light induced degradation of polymeric materials. A small number of polymers have been prepared which contain light sensitive linkages within the polymer chain. Upon exposure, these light sensitive linkages fragment leading to the degradation of the polymer into smaller polymeric units. Alternatively, a separate light absorbing ingredient can generate an active specie (e.g. acid) that can subsequently fragment the polymer (*16*). The degradation of higher molecular weight polymers to lower molecular weight materials can be detected as a change in solubility. Very few practical applications utilizing photodegradation have been commercialized.

Photo/Thermal Reactions. The fifth basic class of photopolymer chemistry that can be used in commercial applications is based more on physical changes in a polymer-based matrix than on chemical reactions. A recent application of this technology is the laser ablation (*17*) of an organic coating on a flat support to directly produce a printing plate. The availability of newer high energy lasers will allow more applications to be based on the photo/thermal mechanism.

Primary Visualization ("Readout") Methods

In the previous section, five basic mechanisms of photopolymer reactions were identified. In order to utilize these photopolymer mechanisms in practical systems, sensitive visualization methods must be used to "develop" the latent images. Six different classes of visualization techniques are used in most commercial applications of photopolymer technology.

Solubility. The most widely used visualization method is based on the detection of changes in solubility initiated through the action of light on photopolymer materials. In many cases the initial photopolymer composition can be dissolved in a solvent while the imaged (exposed) material will be insoluble following "photopolymerization" and "photocrosslinking". Conversely, an initially insoluble material can be made more soluble through various "photodegradation" and "photomolecular reaction" schemes. Some examples of products using this type of readout method include printed wiring board resists, liquid microelectronic resists and printing plates.

Adhesion. The modulation of adhesion with light is used in some important commercial applications of photopolymer technology. The adhesion and cohesion of a photopolymer to other polymeric materials can either be increased or decreased upon the application of light, depending upon the chemistry of the photopolymer formulation and surface properties of the films in the product structure. The development of peel-apart pre-press proofing products, involving no wet chemistry, is an example of the application of adhesion modulation in photopolymer materials (*18-19*).

Color Change. The development of color change within a photopolymer structure has been utilized in a number of commercial products. Specifically, the light induced change of a colorless leuco dye into a colored image forms the basis of several monotone positional pre-press proofing products. Other color change schemes (e.g. photochromic chemistry) have been used in a variety of products.

Phase Change. A number of processes utilizing a change in phase (e.g. liquid to solid) of a photopolymer composition have been commercialized. The recent development of several solid imaging (i.e. stereolithography or rapid prototyping) systems for the direct production of three-dimensional "proofs" from digital data streams (e.g. MRI, magnetic resonance imaging) is a good example of phase change as a readout method for photopolymer reactions. In the most common system, solid imaging consists of the layer-by-layer laser exposure of a liquid photopolymer composition to form a solid, 3-dimensional part representing the original digital image (*20*). The use of high energy lasers to image photopolymer printing plates by an ablation process is also an example of "phase change" as a visualization method.

Refractive Index. The change of refractive index within a photopolymer composition is being used to generate many of the high quality holographic images that are commonly displayed. Photopolymer-based holographic images are formed by employing the difference in the indices of refraction between unpolymerized and polymerized photopolymer material. Other imaging systems also use modulation of refractive index as the visualization method (*21*).

Electrical Conductivity. Changes in the electrical conductivity within or on the surface of a photopolymer film can be used to visualize the results of a photochemical reaction. A limited number of electrophotographic/electrostatic processes have been developed employing this "readout" method (*22*).

Applications of Photopolymer Technology

During the past three decades there has been a rapid increase in the number of commercial applications of photopolymer technology. To facilitate this brief survey of the wide variety of imaging and non-imaging uses of photopolymer chemistry, the practical applications of this technology have been classified into 6 general categories:

- Electronic Materials
- Printing Materials
- Optical and Electro-Optical Materials
- Fabrication of Devices and Materials
- Adhesives and Sealants
- Coatings and Surface Modifications

Additional groupings within each category have been made where appropriate. Every attempt has been made to make this review as inclusive as possible, but because of the extensive nature of this topic, there are many photopolymer applications that have not been included.

Electronic Materials.   Applications of photopolymer technology to electronic products is probably one of the largest imaging categories in terms of sales and total number of patents and publications.   A wide variety of products have been developed for the production of printed wiring boards (PWB) and the manufacture of integrated circuits (IC).

   Manufacture of Printed Wiring Boards.   Printed wiring boards, or printed circuit boards, are usually thin flat panels than contain one or multiple layers of thin copper patterns that interconnect the various electronic components (e.g. integrated circuit chips, connectors, resistors) that are attached to the boards.   These panels are present in almost every consumer electronic product and automobile sold today.   The various photopolymer products used to manufacture the printed wiring boards include film resists, electroless plating resists (23), liquid resists, electrodeposited resists (24), solder masks (25), laser exposed photoresists (26), flexible photoimageable permanent coatings (27) and polyimide interlayer insulator films (28).   Another new use of photopolymer chemistry is the selective formation of conductive patterns in polymers (29).

   Manufacture of Integrated Circuits.   An increasingly important application of photopolymer technology is the use of these materials for the formation of integrated circuits (i.e. chips).   Integrated circuits are multilayer microscale electronic components (e.g. transistors) that are prepared on high purity silicone wafers.   As more and more transistors are incorporated into the small chips for higher performance, the resolution capability of photopolymer materials is continually being improved through the development of novel photopolymer chemistry and manufacturing processes (30).   Some of the photopolymer products used to manufacture integrated circuits include Deep-UV (31-32) and electron-beam resists (33), direct laser imaging materials (34) plus insulating (35) and thick-film pastes (36) that are used to make conductive and resistive patterns on a scale between the extremes of printed wiring boards and integrated circuits.

Printing Materials. The application of photopolymer technology to a wide variety of imaging applications in the printing industry is a quite important.   Most printing plates and pre-press proofing products used today are based on, or utilize some element of, photopolymer chemistry.

   Printing Plates.   Of the many technologies used to manufacture printing plates, photopolymer chemistry is used almost exclusively for the plates that print medium to high quality images.   The most common photopolymer printing applications include lithographic printing plates (37), flexographic plates (38) and new digital laser imageable plates (39).

Pre-Press Proofing Systems. Pre-press proofing products have been developed to provide the printer with a fast accurate color representation of an actual printing job without the expense of preparing the printing plates (a minimum of four plates is required for full color printing) and running the plates on press. Because of the adaptability of photopolymer technology, a number of commercial products have been developed that provide full-color surprint and overlay pre-press proofs (*40*) as well as products for monotone positional proofing applications (*41*). Newer proofing systems based on thermal imaging have also been developed (*42*). A new application of photopolymer chemistry in this area is the development of digital pre-press proofing systems (*43*).

Printing Inks. In addition to printing plates, photopolymer technology has been applied to UV and electron-beam curable printing inks (*44*). These inks have the advantage of rapid cure (thus faster printing speeds) and significantly reduced solvent emissions. UV curable inks have also been developed for other applications including screen stencil (*45*) and ink-jet printing (*46*).

Other Printing Applications. There are a number of lower volume applications of photopolymer chemistry to the printing industry. Some of these include the production of tactile images (raised Braille patterns) (*47*) and screen stencil resists (*48*).

Optical and Electro-Optical Materials. Many of the newest high "value-in-use" applications of photopolymer technology are in the general area of optical and electro-optical materials. Because of the excellent spatial imaging characteristics of photopolymer materials and their ability to rigidify materials to provide aligned molecular structures, photopolymer materials are uniquely positioned to participate in these emerging "high tech" applications.

Liquid Crystal Displays (LCD). Liquid crystal displays, once limited to small devices such as calculators, are now displacing color CRT (cathode ray tube) displays in commercial quantities. The ability to fabricate these display devices at high quality and at low cost is partially due to the wider spread use of photopolymer-based materials. Photopolymer technology is being used for the alignment of liquid crystal (LC) elements (*49*), the orientation of ferroelectric materials (*50*), the synthesis of LC polymers (*51*) and the manufacture of color filters for liquid crystal display applications (*52*).

Nonlinear Optical Materials. Nonlinear optical materials (NLO) rely on the ability of light to interact with specially designed inorganic and organic compounds to provide uniquely modulated light output. Photopolymer materials are being fabricated into various NLO materials and devices that are being used in applications such as optical waveguides (*53*), polarizing optical elements (*54*) and electro-optical switching devices (*55*).

Holographic Materials. One optical application of photopolymer technology that has been receiving a large amount of attention is the area of holographic materials and devices. Holographic elements have the ability to visually reconstruct 3-dimensional images when light is shined on special 2-dimensional structures. Since photopolymer materials have quite high 3-dimensional spatial imaging properties, they are ideally suited for this application. Some of the applications of photopolymer chemistry include the preparation of holographic photopolymer elements (56), reversible holograms (57) and switched holographic gratings (58).

Other Imaging Materials. Photopolymer technology is being utilized in the manufacture of other types of optical elements including rewriteable recording materials (e.g. photochromic imaging) (59) and permanent information recording by processes such as laser ablation.

Fabrication of Devices and Materials. Many of the previously discussed applications of photopolymer technology have utilized photopolymer materials to manufacture a part that is then incorporated into commercial products (e.g. film resists used to manufacture printed circuit boards). One area that has been increasing very rapidly in the past 5-6 years has been the use of photopolymer chemistry to fabricate devices and materials that are used directly in the final product. This development has been made possible, in part, by the design of new photopolymer chemistry that can produce polymerized materials with enhanced physical properties.

3-Dimensional Images and Parts. During many manufacturing processes a prototype or a model is required as a check of the design or fit with another part of the assembly. Until recently many of these prototype parts or molds were sent to a machine shop where it could take up to 3 months, and $20,000 to $40,000, to prepare a single part. If the initial design was not correct, then the entire process would have to be repeated with increasing cost and time delays. Since most of the new part designs, especially in the automotive and aircraft industries, are now created by 3-dimensional computer aided design (CAD) software and are available in a digital format, this has created an opportunity for a new application of photopolymer technology. Several commercial systems have been developed over the past 10 years that can take digital data from CAD software and produce a 3-dimensional part or model using computer driven laser exposure of a liquid photopolymer. This process is referred to by many names including rapid prototyping, solid imaging and stereolithography (60). Other applications of photopolymer chemistry to produce parts and models include micromachining (61), the production of molds using UV curable molding materials (62), the manufacture of optical lenses (63), micro-optical lenses and elements (64), and the preparation (65) and duplication of optical disks (66).

Dental Materials. One very practical application of photopolymer materials is in dental restoratives. Photopolymer chemistry has been adapted for use in composite fillings (67), surface coatings (68) and in the formation of impressions (69).

Medical Applications. Photopolymers are used in variety of medical applications. Some of these applications include the manufacture of implantable bone prosthesis (prepared by rapid prototyping), the use of UV curable splints (*70*) and casts, the utility of orthopedic implants (*71*), the production of contact lenses (*72*), the formation of hydrogels for injectable introocular lenses (*73*) and the manufacture of controlled release hydrogels (*74*). Other biological applications of photopolymer technology include the design of unique materials for cell entrapment (*75*), the production of cell-adhesion-resistant surfaces (*76*) and the immobilization of enzymes (*77*) and micro-organisms.

Membranes. Photopolymer chemistry is being applied to the design and manufacture of a variety of membrane materials. In these applications, photopolymer technology is used to precisely define the microscopic openings in the membrane as it is being formed or to modify an existing membrane. Some of the applications of photopolymer chemistry to membranes include the modification of ultrafiltration membranes (*78*) and the manufacture of amphiphilic (*79*), gas permeable (*80*), untrafiltration (*81*), ion-selective electrode (*82*) and reverse osmosis membranes.

Polymeric Materials. Photopolymer chemistry is also being used to prepare a variety of polymeric materials including crosslinked fibers and films (*83*), microporous gels, microporous polymers (*84*), microparticles (*85*) and battery electrodes (*86*) and components.

Adhesives and Sealants. One specialty application of photopolymer chemistry that has found increasing importance in the past 5 years is in the area of light sensitive adhesives and sealants. The unique ability of photopolymer formulations to harden and tackify or detackify upon exposure to electromagnetic radiation is key to this application. The many examples of the photopolymer technology for the production of these materials include the manufacture of glass laminates (*87*), fiber composites (*88*), inorganic/organic composites (*89*), optical waveguide interconnects (*90*), pressure sensitive adhesives (*91*), hemostatic adhesive (*92*) and high temperature adhesives and sealants (*93*).

Coatings and Surface Modifications. Probably the one application of photopolymer chemistry that has the most worldwide commercial value in terms of product sales is the use of photopolymer materials for curable coatings. Most of the wood paneling and less expensive furniture manufactured today utilize UV or electron-beam curable materials for decorative finishes (e.g. simulation of wood grain) and protective coatings. In addition, the surfaces of many commercially important materials (e.g. textile fibers and polyester films) are being modified by photopolymer processes.

Coatings. The use of photopolymer materials for the preparation of curable coatings is highly diverse. Many of the products used at home and at work have been coated with a photopolymer material for protective and decorative applications. Just a few of the many applications of photopolymer technology include UV curable pigmented (*94*) and clear (*95-96*) protective coatings, protective furniture (*97*) and floor finishes, coatings for biosensors (*98*) and electrodes (*99*), coatings for optical fibers (*100*), a wide variety of release (*101*) and antistatic coatings, insulated foamed coatings (*102*) and thermotropic gel layers for windows (*103*).

Surface Modifications. Basic photopolymer chemistry is also being used for the surface modification of films, textiles fibers, and many other organic-based materials (*104*). Some of the novel applications of photopolymer technology to surface modification include the design of cell repellent treatments and in photografting of various chemical functionality onto the surface of materials to improve color retention, enhance the adhesion of antistatic chemicals or to improve staining resistance.

Conclusion

The commercial applications of photopolymer technology have grown rapidly since the 1950's and 1960's to where they now represent a multi-billion dollar business. The recent developments of novel photopolymer chemistries, materials and exposure systems have been instrumental in the ongoing evolution of various "high tech" products such as holographic materials, optical waveguides, stereolithography systems and high resolution microelectronic resists. The use of photopolymer technology to existing and new applications continues at a rapid pace and many more exciting developments are anticipated in the future.

Acknowledgments. I would like to thank my many colleagues, those within DuPont and those associated with other organizations, for their helpful discussions and thought provoking ideas. In particular, I would like to acknowledge the ongoing support and assistance provided by Mario Grossa, Peter Walker and David Eaton.

Literature Cited

1.   During the 1990-1995 period, the number of distinct references to some aspect of photopolymer technology has exceeded 4,000/year.

2.   Reiser, A. *Photoreactive Polymers*; John Wiley & Sons, New York, 1989, pp 1-21.

3.   Minsk, L. M.; Van Deusen, W. P.; Robertson, E. M. U.S. Patents 2,670,285, 2,670,286, 2,670,287, 1948.

4.   Plambeck, L., Jr. U.S. Patent 2,760,863, 1956.

5.   Personal communications, RadTech Asia '93 Conference, November 10-13, 1993, Tokyo, Japan.

6.   Hageman, H. J. In *Photopolymerisation and Photoimaging Science and Technology*; Allen, N. S., Ed.; Elsevier Applied Science, London, 1989; pp. 1-53.

7.   Pappas, S. P. In *Photopolymerisation and Photoimaging Science and Technology*; Allen, N. S., Ed.; Elsevier Applied Science, London, 1989; pp. 55-73.

8.   Noren, G. K. *Polym. Mater. Sci. Eng.* **1996**, *74*, pp. 321-322.

9.   Dvorchak, M. J. *J. Coat. Technol.* **1995**, *65*, pp. 49-54.
10. Kroggel, M.; Mohr, D.; Mullerhess, W.; Mueller-Hess, W. U.S. Patent 5,206,113, 1993.
11. Barker, I. C.; Allen, N. S.; Edge, M.; Sperry, J. A.; Batten, R. In *Curr. Trends Poly. Photochem.*; Editor Allen, N. S., Horwood, London, 1995; pp 67-80.
12. Liu, J.-H.; Liu, H.-T.; Tsai, F.-R. *Angew. Makromol. Chem.* **1995**, *229*, pp. 63-72.
13. Holme, N. C. R.; Ramanujam, P. S.; Hvilsted, S. *Opt. Lett.* **1996**, *21*, pp. 902-904.
14. Simmons, H. E., III; Hertler, W. R.; Sauer, B. B. *J. Appl. Polym. Sci.* **1994**, *52*, pp. 727-735.
15. Jacobson, R. E. In *Photopolymerisation and Photoimaging Science and Technology*; Allen, N. S., Ed.; Elsevier Applied Science, London, 1989; pp. 149-186.
16. Hiro, M.; Frechet, J. M. J. *ACS Symp. Ser.* **1996**, *620*, pp. 381-386.
17. Holtz, S.; Bargon, *J. Appl. Phys. A: Mater. Sci. Process.* **1995**, *A60*, pp. 529-535.
18. Grossa, M.; Sondergeld, M. EPO Published Patent Application 640,879, 1995.
19. Choi, H.-C.; Kniazzeh, A.; Habbal, F. *Mater. Res. Soc. Symp. Proc.* **1996**, *409*, pp. 397-402.
20. Burns, M. *Automated Fabrication*; PTR Prentice Hall, Englewood Cliffs, New Jersey, 1993.
21. Bach, H.; Anderle, K.; Fuhrmann, T.; Wendorff, J. H. *J. Phys. Chem.* **1996**, *100*, pp. 4135-4140.
22. Yoshida, A.; Goto, K.; Ooomori, H. Morikawa, Y. JP Published Patent Application 80/62,942, 1996.
23. Sugihara, S. *Nippon Setchaku Gakkaishi* **1995**, *31*, pp. 12-20; *Chem. Abstr.* **1995**, *123*, 128944.
24. Merricks, D. In *Spec. Polym. Electron. Optoelectron*; Chilton, J. A.; Goosey, M.T., Eds.; Chapman & Hall, London, **1995**, pp. 37-80.
25. Matynia, T.; Kutyla, R.; Bukat, K.; Pienkowska, B. *J. Appl. Polym. Sci.* **1995**, *55*, pp. 1583-1588.
26. Decker, C.; Elzaouk, B. *Polym. Mater. Sci. Eng.* **1995**, *72*, pp. 6-7.
27. Dueber, T. E.; Schadt, F. L., III EPO Published Patent Application 676,669, 1995.
28. Yoshikawa, M. JP Published Patent Application 72/83,118, 1995.
29. Dai, L.; Griesser, H. J.; Hong, X.; Mau, A. W. H.; Spurling, T. H.; Yang, Y.; White, J. W. *Macromolecules* **1996**, *29*, pp. 282-287.
30. Reichmanis, E.; Nalamasu, O.; Novembre, A. E. *J. Photopolym. Sci. Technol.* **1995**, *8*, pp. 709-728.
31. Ogawa, T.; Oizumi, H.; Ito, M.; Saitou, N. *Microelectron. Eng.* **1996**, *30*, pp. 287-290.
32. Paniez, P. J.; Pain, L. *J. Photopolym. Sci. Technol.* **1995**, *8*, pp. 643-652.
33. Crivello, J. V.; Shim, S.-Y. *Chem. Mater.* **1996**, *8*, pp. 376-381.
34. Decker, C.; Elzaouk, B. Polym. Mater. Sci. Eng. 1995, 72, pp. 6-7.

35. Masaki, Y.; Iwanaga, K.; Yoshimura, A. JP Published Patent Application 70/45,120, 1995.
36. Hayakawa, K.; Inaba, A.; Kuno, H.; Matsuno, H.; Terii, A. S.; Jeromu, D. S. JP Published Patent Application 81/86,005, 1996.
37. Takizawa, K. *Fuji Film Res. Dev.* **1995**, *40*, pp. 39-43; *Chem. Abstr.* **1995**, *123*, 97689.
38. Cusdin, G. *Tappi Journal* **1995**, *4*, pp. 177-182
39. Konishi, K. *Techno Cosmos* **1995**, *7*, pp. 9-14.
40. Bruno, M. H. *Principles of Color Proofing*, GAMA Communications, Salem, NH, 1986, 133-180.
41. Sheets, T. M. U.S. Patent 4,622,286, 1986.
42. Bloom, I. B. K.; Minns, R. A. U.S. Patent 5,387,479, 1995.
43. *Res. Discl.* **1996**, *385*, pp. 298-299.
44. Holman, R. *Eur. Coat. J.* **1995**, pp. 610-612.
45. Rose, J. *Screen Disp.* **1995**, pp. 28-29.
46. Kiyota, K.; Sekioka, C.; Sugie, M. WPO Published Patent Application 96/28,518, 1996.
47. Roland, G. U.S. Patent 5,397,683, 1995.
48. Gybin, A. S.; Johnson, K. K.; Komatsu, T.; Van Iseghem, L. C. WPO Published Patent Application 95/12148, 1995.
49. Akiyama, H.; Momose, M.; Ichimura, K.; Yamamura, S. *Macromolecules* **1995**, *28*, pp. 288-293.
50. Kelly, S. M. *J. Mater. Chem.* **1995**, *5*, pp. 2047-2061.
51. Lub, J.; Broer, D. J.; Hikmet, R. A. M.; Nierop, K. G. J. *Liq. Cryst.* **1995**, *18*, pp. 319-326.
52. Imamura, T.; Mikami, S.; Sumiyoshi, I.; Tsushima, H; Watanabe, E. EPO Published Patent Application 725,315, 1996.
53. Weber, A. M. *Polym. Mater. Sci. Eng.* **1995**, *72*, p. 65.
54. Barachevsky, V. A. *Proc. SPIE-Int. Soc. Opt. Eng.* **1995**, *2208*, pp. 184-195.
55. Natarajan, L. V.; Sutherland, R. L.; Tondiglia, V. P.; Bunning, T. J. *J. Nonlinear Opt. Phys. Mater.* **1996**, *5*, pp. 89-98.
56. Steijn, K .W. *Proc. SPIE-Int. Soc. Opt. Eng.* **1996**, *2688*, pp. 123-134.
57. Ghailane, F.; Manivannan, G.; Lessard, R. A. *Opt. Eng.* **1995**, *34*, pp. 480-485.
58. Sponsler, M. B. *J. Phys. Chem.* 1995, *99*, pp. 9430-9436.
59. Yamaoka, T.; Tamaoki, N. *Senryo to Yakuhin* **1995**, *40*, pp, 61-76; *Chem. Abstr.* **1995**, *123*, 183103.
60. Petillon, N.; Jezequel, J. Y.; Andre, J. C. *J. Imaging Sci. Technol.* **1996**, *40*, pp. 42-49.
61. Loechel, B.; Maciossek, A.; Quenzer, H. J.; Wagner, B. *J. Electrochem. Soc.* **1996**, *143*, pp. 237-244.
62. Fifield, C. C. U.S. Patent 5,381,735, 1995.
63. Hirano, N.; Kazama, H.; Matsuoka, S. JP Published Patent Application 70/3,025, 1995.
64. Eisenber, N. P.; Manevich, M.; Klebanov, M.; Shutina, S.; Lyubin, V. *Proc. SPIE-Int. Soc. Opt. Eng.* **1995**, *2426*, pp. 235-241.

65.   Kraakman, P. A.; Morton, R. D.; Put, P. L. M.   WPO Published Patent Application 95/34,894, 1995

66.   Yoshida, K. JP Published Patent Application 81/61,774, 1996.

67.   Wakasa, K.; Chowdhury, N. A..; Priyawan, R.; Uoshida, Y.; Ikeda, A.; Hirose, T.; Yamaki, M. *J. Mater. Sci. Lett.* **1996**, *15*, pp. 134-136.

68.   Goebel, R. German Published Patent Application 195,09,289, 1996.

69.   Moszner, N.;   Rheinberger, V.; Salz, U. EPO Published Patent Application 634,393, 1995.

70.   Ottl, P.; Schmitz, C.; Janda, R. Weigl, P. *Dtsch. Zahnaerztl. Z.* **1995**, *50*, pp. 471-474.

71.   Anseth, K. S.; Shastri, V. R.; Laurencin, C. T.; Langer, R. *Polym. Mater. Sci. Eng.* **1996**, *74*, pp. 385-386.

72.   Mueller, B. WPO Published Patent Application 96/24,077, 1996.

73.   Tamada, Y.; Shigemoto, Y.; Yasuda, K.; Sawa, M.; Masuda, K.; Ichimura, K.; Yamauchi, A. *Seitai Zairyo* **1995**, *13*, pp. 117-122; *Chem. Abstr.* 1995, *123*, 123040.

74.   Hubbell, J. A.; Pathak, C. P.; Sawhney, A. S.; Desai, N. P.; Hill, J. L. U.S. Patent 5,410,016, 1995.

75.   Hertzberg, S.; Moen, E.; Vogelsang, C.; Oestgaard, K. *Appl. Microbiol. Biotechnol.* **1995**, *43*, pp. 10-17.

76.   Drumheller, P. D.; Hyubbell, J. A. *J. Biomed Mater. Res.* **1995**, *29*, pp. 207-215.

77.   Uhlich, T.; Ulbricht, M.; Tomaschewski, G. *Enzyme Microb. Technol.* **1996**, *19*, pp. 124-131.

78.   Yamagishi, H. Crivello, J. V.; Belfort, G. *J. Membr. Sci.* **1995**, *105*, pp. 237-247.

79.   Liu, J.-H.; Chung, Y.-C.; Lin, M.-T. *J. Appl. Polym. Sci.* **1995**, *55*, pp. 1441-1449.

80.   Matsui, S.; Nakagawa, T. *J. Photopolym. Sci. Technol.* 1995, *8*, pp. 321-324.

81.   Yamagishi, H.; Cruvekki, J. V.; Belfort, G. *J. Membr. Sci.* 1995, *105*, pp. 249-259.

82.   Bratov, A.; Abramova, N.; Munoz, J.; Dominquez, C.; Alegret, S.; Bartroli, J. *Anal. Chem.* **1995**, *67*, pp. 3589-3595.

83.   Allen, N. S.; Hurley, J. P.; Bannister, D.; Follows, G. W. *Eur. Polym. J.* **1992**, *28*, pp. 1309-1314.

84.   Chew, C. H.; Ng. S. C.; Gan, L. M.; Teo, W. K.; Gu, J. Y.; Zhang, G. Y. *J. Macromol. Sci., Pure Appl. Chem.* **1995**, *A32*, pp. 969-980.

85.   Katsuta, T.; Aotani, S. JP Published Patent Application 70/10,912, 1995.

86.   Allcock, H. R.; Nelson, C. J.; Coggio, W. D. U.S. Patent 5,414,025, 1995.

87.   Shi, W.; Ranby. B. *J. Appl. Polym. Sci.* **1996**, *59*, pp. 1951-1956.

88.   Coons, L. S.; Rangarajan, B.; Scranton, A. B. *Polym. Mater. Sci. Eng.* **1996**, *74*, pp. 389-390.

89.   Sellinger, A.; Laine, R. M. *Macromolecules* **1996**, *29*, pp. 2327-2330.

90. Yardley, J. T.; Eldada, L.; Norwood, R. A., Stengel, K. M. T.; Shacklette, L. W.; Wu, C.; Xu, C. *MCLC S&T, Sect. B: Nonlinear Opt.* **1996**, *15*, pp. 443-450.

91. Chandran, R.; Ramharack, R. U.S. Patent 5,391,406, 1995.

92. Nakayama, Y.; Matsuda, T. *Asaio J.* **1995**, *41*, pp. M374-M378: *Chem. Abstr.* **1996**, *124*, 144885.

93. Woods, J.; Masterson, M.; McArdle, C.; Burke, J. *Polym. Mater. Sci. Eng.* **1996**, *74*, pp. 317-318.

94. Howell, B. F.; De Raaff, A.; Marino, T. *Polym. Mater. Sci. Eng.* **1996**, *74*, pp. 387-388.

95. Decker, C. *Adv. Chem. Ser.* **1996**, *249*, pp. 319-334.

96. Valet, A. *Farbe Lack* **1996**, *102*, pp. 40-42, 44-46.

97. Salthammer, T. *J. Coat. Technol.* **1996**, *68*, pp. 41-47.

98. Leca, B.; Morelis, R. M.; Coulet, P. R. *Mikrochim. Acta* **1995**, *121*, pp. 147-154.

99. Yoshimura, E.; Tezuka, S. EP Patent Application 676,636, 1995.

100. Murray, K. P.; Szum, D. M.; Zimmerman, J. M. WPO Published Patent Application 96/23,828, 1996.

101. Priou, C.; Soldat, A.; Cavezzan, J.; Castellanos, F.; Fouassier, J. P. *J. Coat. Technol.* **1995**, *67*, pp. 71-78.

102. Suga, M.; Konishi, S.; Takahata, N.; Kato, Y.; Suzuki, Y.; Okuie, H. JP Published Patent Application 70/45,145, 1995.

103. Jahns, E.; Kroner, H.; Mielke, M. EPO Published Patent Application 678,534, 1995.

104. Chan, C.-M.; Ko, T.-M.; Hiraoka, H. *Surf. Sci. Rep.* **1996**, *24*, pp. 1-54.

ACRYLATE SYSTEMS

Chapter 2

Photopolymerization and Electrooptic Properties of Polymer Network/Ferroelectric Liquid-Crystal Composites

C. Allan Guymon, Lisa A. Dougan, and Christopher N. Bowman[1]

Department of Chemical Engineering, University of Colorado, Campus Box 424, Boulder, CO 80309–0424

Polymer/liquid crystal (LC) composites have recently been the focus of considerable attention. One group of these composites, namely polymer stabilized ferroelectric liquid crystals (PSFLCs), show great potential for display technology due to the inherently fast switching speed of the liquid crystal and the mechanical strength imparted by the polymer. This study examines the effects of introducing polymer networks into a ferroelectric liquid crystal (FLC) and the polymerization behavior of cross-linking monomers in the ordered media of an FLC. Phase behavior changes significantly after addition of monomer, but upon polymerization the phase transitions return close those of the pure FLC. On the other hand, electro-optic properties of the system, such as the ferroelectric polarization, are changed significantly by introducing a polymer network. These changes are highly dependent on the monomer structure and the LC phase in which polymerization is initiated. The polymerization behavior is also highly dependent on the order of the media. As the order of the LC phase increases, the rate of polymerization increases significantly for both mesogenic and non-mesogenic monomers. This behavior occurs as the monomers segregate within the LC, increasing the local monomer concentration and thus the polymerization rate.

Liquid crystals (LCs) have been the focus of considerable research for many years and have been developed for use in a wide array of applications. Recently, the development and application of polymer/LC composites has become an area of great interest in LC research. Introducing polymers in LC systems increases the inherent mechanical strength and may dramatically change the LC phase behavior and electro-optic properties (*1*). Conversely, the directional ordering present in liquid crystals forms a fascinating media in which to study polymerizations (*2*).

[1]Corresponding author.

To produce novel LC phase behavior and properties, a variety of polymer/LC composites have been developed. These include systems which employ liquid crystal polymers (*3*), phase separation of LC droplets in polymer dispersed liquid crystals (PDLCs) (*4*), incorporating both nematic (*5,6*) and ferroelectric liquid crystals (*6-10*). Polymer/LC gels have also been studied which are formed by the polymerization of small amounts of monomer solutes in a liquid crystalline solvent (*11*). The polymer/LC gel systems are of particular interest, rendering bistable chiral nematic devices (*12*) and polymer stabilized ferroelectric liquid crystals (PSFLCs) (*1,13*), which combine fast electro-optic response (*14*) with the increased mechanical stabilization imparted by the polymer (*15*).

The influence of the liquid crystalline solvent on polymerization during the formation of polymer/LC composites is also of great import. The polymerization behavior may change dramatically when performed in LC phases (*16*). The polymer morphology is highly dependent on the polymerization conditions and consequently the interactions between the polymer and liquid crystal may also change considerably, thus changing the electro-optic properties. The properties of the polymer/LC composite will therefore be greatly influenced by the polymerization behavior. Many studies have examined the polymerization of oriented LC monomers to produce LC polymers (*17*) including research specifically examining the effect of the LC phase on polymerization (*3,16,18-27*). The anisotropy exhibited in the LC phases has been found to enhance dramatically the polymerization rate for both monoacrylates (*18,22*) and diacrylates (*20*) while the increased rates are strongly dependent on the temperature and LC phase of polymerization (*19*).

The reasons behind this accelerated rate behavior have been attributed to a decrease in chain transfer processes (*28,29*) and a decreased termination rate (*24,25*) indicated by molecular weight measurements (*26*). Recently, direct evidence of decreases in the termination rate have been shown (*27*) and in these studies both the termination and propagation kinetic constants were determined for polymerizations exhibiting enhanced rates in a smectic phase. The propagation constant, k_p, decreases slightly in the ordered phase from the isotropic polymerization. Such a decrease would be expected because of the lower temperature in the smectic phase. The termination kinetic constant, k_t, however, decreases almost two orders of magnitude for the ordered polymerization, indicating a dramatically suppressed termination rate.

Similar behavior has also been observed for polymerizations of small amounts of various monomers in ordered LC phases (*30*). The rate of polymerization is enhanced considerably for a non-mesogenic diacrylate in the smectic C* phase and is more than three times that observed in the isotropic phase of the same LC and over six times that observed for polymerization in an isotropic solvent. Similar results were observed for a variety of mesogenic and other non-mesogenic monomers (*31*). Interestingly, the mechanisms behind this rate enhancement is not the same for all monomers and is highly dependent on the segregation behavior.

This study examines the effects of the polymerization conditions on the electro-optic performance of PSFLCs, and the influence of the LC ordering on the polymerization behavior of various monomers is discussed. Basic electro-optic properties, such as ferroelectric polarization and optical response time, and phase behavior were determined for systems with increasing monomer concentration as well as for systems polymerized at different temperatures corresponding to different LC phases. Additionally, the polymerization behavior of different amorphous and LC diacrylate monomers dissolved in an FLC matrix is discussed for polymerizations in different LC phases and at different temperatures.

Experimental Section

The amorphous diacrylate monomers chosen for study were two commercially available monomers, *p*-phenylene diacrylate (PPDA) and 1,6-hexanediol diacrylate (HDDA) (Polysciences, Inc., Warrington, PA). The liquid crystalline diacrylate studied was 1,4-di-(4-(6-acryloyloxyhexyloxy)benzoyloxy)-2-methylbenzene (C6M) (*13*). Chemical structures of these monomers as well as pertinent physical and LC properties are given in Figure 1. All monomers were used without further purification. The ferroelectric liquid crystal mixture consisted of a 1:1 mixture of W7 and W82 (*1*) (Displaytech, Boulder, CO). This mixture exhibits isotropic (I), smectic A

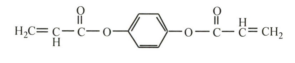

(a)

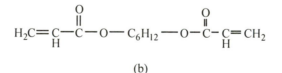

(b)

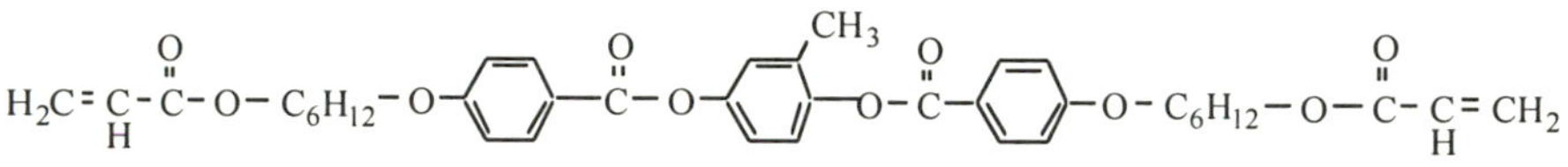

$H_2C = C - C - O - C_6H_{12} - O - \bigcirc - C - O - \bigcirc - O - C - \bigcirc - O - C_6H_{12} - O - C - C = CH_2$

(c)

FIGURE 1. Chemical structures of the monomers used in this work. Shown are a) *p*-phenylene diacrylate (PPDA- melting point: 89° C), b) 1,6-hexanediol diacrylate (HDDA- melting point: 5° C, boiling point: 316° C), and c) C6M, a liquid crystalline diacrylate (phase sequence: Isotropic → 116° C → Nematic → 86° C → Crystalline.)

(SA), smectic C* (SC*) and crystal (X) phases (I ↔ 58°C ↔ SA* ↔ 48°C ↔ SC* ↔ 13°C ↔ X]. All photopolymerizations were initiated with Irgacure 907 (2-methyl -1-[4-(methylthio)phenyl]-2-(4-morpholinyl)-1-propanone, Ciba-Geigy, Hawthorne, NY).

Phase transition temperatures of the monomer/polymer/LC mixtures were determined by first heating the samples to the isotropic state and then cooling slowly at approximately 1° C per minute. The clearing point (Isotropic → Smectic A) and the Smectic A → Smectic C* transition were found by observing optical changes and switching behavior. These phase transitions were confirmed using differential scanning calorimetry.

Ferroelectric polarization values were obtained using an automated polarization tester (APT, Displaytech) by applying a 6 V/μm electrical field across a 4μm rubbed polyimide indium tin oxide cell (Standish, Lake Mills WI) and integrating the induced current peak (*32*). To determine optical response time, a square wave electric field (6 V/μm) was then applied across the 4μm cell. HeNe laser light (10mW) at 630 nm was passed through the cell between crossed polarizers, and its intensity was determined by an optical intensity sensor. The optical response time was then found using a digitizing oscilloscope. All measurements and observations were determined using surface stabilized samples (*33*).

Photopolymerizations were monitored using a differential scanning calorimeter modified with a photocalorimetric accessory (Perkin-Elmer DSC-DPA 7). The photocalorimetric accessory included a monochromator to produce light of a specific wavelength and monochromatic light with a wavelength of 365 nm was selected for these studies. Polymerizations were initiated using a light intensity of 2.5 mW/cm^2. The DSC cell was purged with nitrogen for 10 minutes prior to polymerization and throughout the reaction to prevent oxygen inhibition.

Samples for infra-red absorption measurement were introduced between two rubbed nylon coated calcium fluoride substrates spaced 10 μm apart. To insure proper parallel alignment, samples were cooled at 0.02° C/minute from the isotropic to the Smectic C* phase. The alignment was then checked using polarizing microscopy. Polarized IR spectra (32 scans per spectrum) were obtained using an FTIR spectrometer (IFS-66; Bruker, Pillerica, MA) equipped with a wire grid polarizer at a resolution of 2 cm^{-1}.

Results and Discussion

Phase Behavior and Electro-optic Properties. With the great potential of PSFLCs, it is important to understand the changes induced by the polymer network on the FLC properties. In order to do so, two different non-mesogenic monomers, HDDA and PPDA, have been used. The structure of these monomers is quite similar (See Figure 1) with the only difference being that the phenyl group in PPDA is replaced with a six carbon alkyl chain for HDDA. These monomers, despite their structural similarity, have much different physical properties and consequently the polymers formed from HDDA and PPDA may also influence the FLC phase and electro-optic behavior differently.

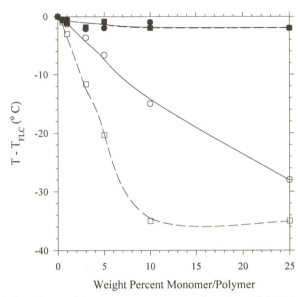

FIGURE 2. Reduced transition temperatures as a function of PPDA monomer and polymer concentration for the isotropic to smectic A transition before (○) and after polymerization (●) and for the smectic A to smectic C* transition transition before (□) and after polymerization (■).

The phase behavior is changed considerably upon addition of PPDA monomer to the FLC as shown in Figure 2. The reduced transition temperatures for the LC phases, i.e. the transition temperature for the pure FLC subtracted from that of the FLC/monomer (or polymer) mixtures, are plotted as a function of the concentration before and after polymerization. Before polymerization the reduced transition temperatures decrease almost linearly for the first order isotropic to smectic A transition, as would be expected. The reduced temperatures for the transition from the smectic A to the smectic C* phase for the monomer/FLC mixtures also decrease linearly with concentration, but the decrease is considerably more pronounced. This decrease continues until the LC is saturated in monomer (about 13 wt%).

After polymerizing, the phase behavior changes dramatically. The phase transition temperatures return to values very close to those observed in the pure FLC. The interactions which lower the transitions in monomer/FLC systems are not significant in polymer/FLC systems. Similar results are observed in HDDA/FLC systems. The only notable exception is that the monomer saturation concentration occurs at a significantly lower concentration (5 wt%).

Although phase behavior is comparable for both HDDA and PPDA monomer mixtures, the electro-optic properties may behave quite differently. It has been shown previously that the polymer can considerably change PSFLC properties (1,34), but these changes are not the same for all polymers. To investigate the effects of different polymer networks on the FLC mixture, the ferroelectric polarization density, or net macroscopic dipole, of both 2% PPDA and HDDA polymer/W7,W82 samples was measured in the smectic C* phase. Both polymers affect the polarization differently as shown in Figure 3. At lower temperatures the 2% HDDA polymer sample exhibits polarization values close to those in the FLC. As the temperature is increased, the polarization drops considerably until temperatures close to the transition are attained. At this point, the values have decreased more than a factor of three from those at the same temperature in the FLC. On the other hand, 2% PPDA polymer samples behave much differently. The polarization values are lower than both the 2% HDDA system and the pure FLC. When the sample is heated, the polarization still decreases but the magnitude of decrease is significantly lower than that observed in the HDDA polymer mixtures. As indicated by these results, the differences in the PPDA and HDDA polymers result in marked differences in electro-optic properties.

Further evidence contrasting the changes induced by introduction of different polymer networks is illustrated in Figure 4. The ferroelectric polarization is given as a function of the temperature at which the sample is polymerized for both 2% PPDA and HDDA polymer in W7,W82. All of the data point were measured in the smectic C* phase at 35° C; however, each samples was polymerized at different temperatures as indicated in the figure. The temperature ranges for the LC phases in which polymerization occurred are also delineated. As is apparent from the figure, the dependence of polarization on polymerization temperature is relatively insignificant for PPDA/FLC composites. Values deviate only slightly over the 50° C range of polymerization temperature and have little if any connection to the LC phase in which polymerization was initiated.

Results for HDDA, however, are in striking contrast. The polarization jumps over fifty percent when the polymerization temperature is changed by only 25° C. The values also increase steadily as the polymerization temperature is increased within the smectic C* phase. These values then seem to level out in smectic A and drop as the polymerization temperature is raised in the isotropic phase. Similar trends are also observed in the optical response time as the polymerization temperature is changed (*29*). Some of the reasons behind this behavior can be understood upon using polarizing microscopy to examine the HDDA PSFLC systems. HDDA/FLC samples polymerized in the Smectic C* phase show relatively small, well oriented domains. As the polymerization temperature increases, the domains increase in size and the orientation of the domains becomes more random. For isotropic polymerizations the domains are large, but their orientation is extremely random and large defects are observed. This behavior indicates that no order is imparted on the polymer during polymerization. Thus, the polymer surface effects on the FLC are fairly significant and do not allow the surface layer to align the FLC molecules. Therefore, the net macroscopic dipole, or spontaneous polarization, decreases dramatically when the PSFLC is polymerized in the isotropic state. These results indicate that the LC phase, and thus the order of the polymerization media, dramatically influences the electro-optic behavior of the system for HDDA/FLC composites.

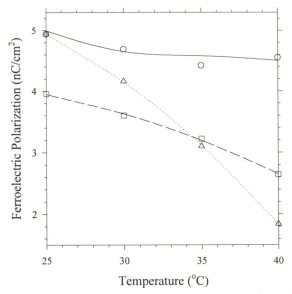

FIGURE 3. The ferroelectric polarization versus temperature for W82,W7 (O), 2% PPDA polymer (□) and 2% HDDA polymer (△) in W82,W7.

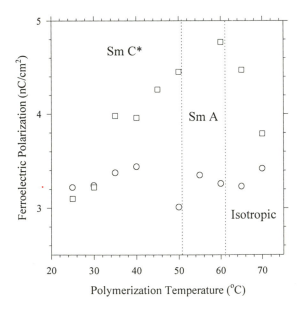

FIGURE 4. Ferroelectric polarization as a function of polymerization temperature for 2% PPDA (O) and 2% HDDA (□) in W82,W7. LC phases in which polymerizations were initiated are also denoted.

Polymerization Behavior and Monomer Segregation. The differences in electro-optic properties for the two different composites suggest that the interactions between the FLC and polymer are significantly different for different polymer systems. This feature implies that the LC significantly influences the formation of the polymers and the consequent polymer morphology, indicating that the polymerization behavior may also be different than would be expected in typical polymerizations. In fact, for small amounts of certain diacrylates in an FLC, the rate for polymerizations in ordered smectic phases is dramatically higher than that observed in the isotropic phase (*24*). To understand further the effects of the FLC phase on these rate differences and to provide more insight into the electro-optic behavior, the polymerization rate was determined for HDDA in different LC phases and at different temperatures. Figure 5 shows the normalized polymerization rate as a function of double bond conversion for 5% HDDA in W7,W82. The lowest temperature corresponds to a smectic C* phase polymerization, the highest temperature to an isotropic polymerization and the mid-range temperature to a smectic A polymerization. The polymerization in the ordered phases is dramatically different than the isotropic polymerization. The maximum rate for the smectic C* polymerization is almost three times that observed in the isotropic polymerization. The acceleration is not as pronounced in the Smectic A phase, but the rate is still significantly greater than in the isotropic polymerization. The additional conversion in the Smectic A polymerization is an artifact caused by an exothermic phase transition during the polymerization (*24*). Therefore, as the

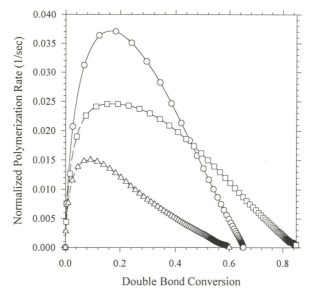

FIGURE 5. Photopolymerization rate as a function of double bond conversion for 5% HDDA in W82,W7 at 23° C - Smectic C* (○), at 50° C - Smectic A (□), and at 65° C - Isotropic (△).

polymerization temperature decreases, and consequently the order of the LC increases, an increase in the polymerization rate is observed. Similar results are observed in PPDA polymerizations. These results suggest that the order of the LC
influences the monomer behavior and may act to segregate the monomer.

To determine if this phenomenon is isolated to amorphous monomers, a liquid crystalline diacrylate (C6M) was polymerized in W7,W82 at temperatures corresponding to the two smectic phases as well as the isotropic phase. The polymerization rate for C6M is plotted as a function of time for representative temperatures
in Figure 6. Again, the polymerization shows marked acceleration in the ordered
smectic C* phase and occurs much faster than the isotropic polymerization. As
seen in the HDDA polymerizations, the smectic A rate also lies between the rates of
the other two polymerization temperatures.

Therefore, for three monomers with different chemical structures and varying
physical and mesogenic properties, the polymerization rate behavior is similar. The
acceleration of the rate in the ordered phases is highly indicative of segregation of the
monomers in the ordered phases of the FLC. To determine if segregation is actually
occurring in these systems, polarized infra-red (IR) spectroscopy experiments were
performed on the monomer/W7,W82 systems. Figure 7 shows a polar plot of the
absorbance of the C=C stretch (at 1635 cm^{-1}) scaled by the minimum observed absorbance plotted versus the polarization angle (Ω) for HDDA, PPDA and C6M
monomer/FLC mixtures. If the monomers were isotropically distributed or randomly segregated throughout the sample, the absorbance would be the same for all
polarization angles resulting in a circular curve on the polar plot. Conversely, if the
absorbance changes with polarization angle, then some degree of directional segregation and orientation is present. Such is the case for all three monomers. The absorbance for the HDDA system exhibits maxima at 0° and 180°, corresponding to light
polarized parallel to the smectic layers. The absorbance for light polarized parallel
to the smectic layers is over three times that for perpendicularly polarized light.
Such behavior would be consistent with the reactive double bonds in the HDDA
molecules segregating between the smectic layers. If the whole HDDA molecule is in
fact segregating swelling in the smectic layers would be observed. These results are,
in fact, observed, and the layer spacing increase is approximately that which would
be expected if all of the molecules segregate between the layers (*30*). On the other
hand, the IR maxima and minima for the double bonds in C6M and PPDA are the
opposite of those observed in HDDA and no smectic layer spacing increase is observed by X-ray diffraction indicating that the double bonds preferentially align perpendicular to the smectic layers and parallel to the FLC molecules themselves. This
behavior would be consistent with the C6M and PPDA molecules aligning with the
FLC molecules, thus reducing the volume in which the double bonds are located and
increasing the local concentration and consequently accelerating the polymerization
rate.

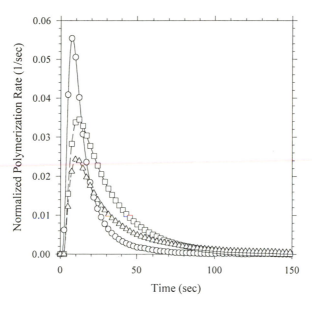

FIGURE 6. Photopolymerization rate versus time for 6% C6M in W82,W7 at 23° C- Smectic C* (○), at 54° C- Smectic A (□), and at 65° C- Isotropic (△).

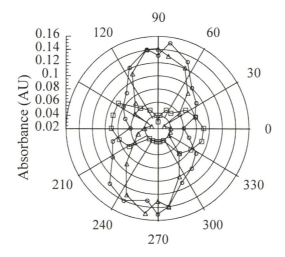

FIGURE 7. Polar plot of the absorbance of the IR band at 1635 cm^{-1} for the acrylate C=C stretch for 15% C6M (○), 5% HDDA (□), and 5% PPDA (△).

Conclusions

The introduction of a polymer network into an FLC dramatically changes phase and electro-optic behavior. Upon addition of monomer to the FLC, the phase transitions decrease and after polymerization return to values close to that observed in the neat FLC. The phase behavior is similar for the amorphous monomers, HDDA and PPDA. The electro-optic properties, on the other hand, are highly dependent on the monomer used to form the polymer/FLC composite. The ferroelectric polarization decreases for both HDDA and PPDA/FLC systems, but the values for each show extremely different temperature dependence. Further evidence illustrating the different effects of each of the two polymers is found upon examining the polarization as both the temperature and LC phase of polymerization are changed. In PPDA systems the polarization remains fairly independent of the polymerization temperature. On the other hand, the polarization increases steadily as the polymerization temperature of HDDA systems is increased in the ordered LC phases.

The polymerization temperature and LC phase also have a dramatic effect on the polymerization behavior in these polymer/FLC composites. For both mesogenic and non-mesogenic monomers the polymerization rate increases considerably as the order of the LC phase increases and as temperature decreases. Maximum rates in the smectic C* phase are two to three times that observed in the isotropic phase. Rate acceleration is also observed in the smectic A phase. The rate acceleration can be attributed to orientation and segregation of the reactive double bonds. Although the pattern of segregation is different for mesogenic and non-mesogenic monomers, the segregation for both types of monomer results in a reduced volume in which the reactive double bonds are found, thus increasing the local concentration and the rate.

Acknowledgments

The authors would like to thank the Department of Defense for its support of this research through a fellowship to CAG and Dr. Dirk Broer for providing the C6M monomer. The financial support of the National Science Foundation (MRG Grant # DMR-9224168 and CTS-9453369) is also gratefully acknowledged.

Literature Cited

(1)   Guymon, C. A.; Hoggan, E. N.; Walba, D. M.; Clark, N. A.; Bowman, C. N., *Liq. Crystals* **1995**, *19*, 719.
(2)   Hoyle, C. E.; Watanabe, T.; Whitehead, J. B. *Macromolecules* **1994**, *27*, 6581.
(3)   Broer, D. J.; Finkelmann, H.; Kondo, K. *Makromol. Chem.* **1988**, *189*, 185.
(4)   Drzaic, P. S. *Liquid Crystal Dispersions*; World Scientific: Singapore, 1995.
(5)   Braun, D.; Frick, G.; Grell, M.; Klines, M.; Wendorff, J. H. *Liq. Cryst.* **1992**, *11*, 929.
(6)   Kitzerow, H.-S.; Molsen, H.; Heppke, G. *Appl. Phys. Lett.* **1992**, *60*, 3093.

(7) Molsen, H.; Kitzerow, H.-S. *J. Appl. Phys.* **1994**, *75*, 710.

(8) Kitzerow, H.-S. *Liq. Cryst.* **1994**, *16*, 1.

(9) Zyryanov, V. Y.; Smorgon, S. L.; Shabanov, V. F. **1993**, *143*, 271.

(10) Lee, K.; Suh, S.-W.; Lee, S.-D. *Appl. Phys. Lett.* **1994**, *64*, 718.

(11) Hikmet, R. A. M.; Zwerver, B. H., *Liq. Crystals* **1991**, *10*, 835.

(12) Yang, D. K.; Chien, L. C.; Doane, J. W., *Appl. Phys. Lett.* **1992**, *60*, 3102.

(13) Hikmet, R. A. M.; Boots, H. M. J.; Michielsen, M. *Liq. Crystals* **1995**, *19*, 65.

(14) Walba, D. M. *Science* **1995**, *270*, 250.

(15) Lester, G.; Coles, H.; Murayama, A.; Ishikawa, M. *Ferroelectrics* **1993**, *148*, 389.

(16) Hoyle, C. E.; Griffin, A. C.; Kang, D.; Chawla, C. P. In *Irradiation of Polymeric Materials*; ACS Symposium Series 527; Reichmanis, E., O'Donnell, J. H., Frank, C. W., Eds.; American Chemical Society: Washington, D.C., 1993; pp 118-130.

(17) Barrall, E. M., II; Johnson, J. F. *J. Macromol. Sci. Rev. Macromol. Chem.* **1979**, *17*, 137.

(18) Broer, D. J.; Mol, B. N. *Makromol. Chem.* **1989**, *190*, 19.

(19) Broer, D. J.; Boven, J.; Mol, G. N.; Challa, G. *Makromol. Chem.* **1989**, *190*, 2255.

(20) Broer, D. J.; Hikmet, R. A. M.; Challa, G. *Makromol. Chem.* **1989**, *190*, 3201

(21) Broer, D. J.; Mol, G. N.; Challa, G. *Makromol. Chem.* **1991**, *192*, 59.

(22) Hoyle, C. E.; Kang, D.; Jariwala, C.; Griffin, A. C. *Polymer* **1993**, *34*, 3070.

(23) Hoyle, C. E.; Chawla, C. P.; Griffin, A. C. *Polymer* **1989**, *60*, 1909.

(24) Hoyle, C. E.; Kang, D. *Macromolecules* **1993**, *26*, 844.

(25) Hoyle, C. E.; Chawla, C. P.; Kang, D.; Griffin, A. C. *Macromolecules* **1993**, *26*, 758.

(26) Hoyle, C. E.; Chawla, C. P. *Macromolecules* **1995**, *28*, 1946.

(27) Hoyle, C. E.; Watanabe, T. *Macromolecules* **1994**, *27*, 3790.

(28) Heynderickx, I.; Broer, D. J.; Van Den Boom, H.; Teesselink, W. J. D. *J. Polym. Sci., Polymer Phys. Ed.* **1992**, *30*, 215.

(29) Hikmet, R. A. M.; Lub, J.; Tol, A. J. W. *Macromolecules* **1995**, *28*, 3313.

(30) Guymon, C. A.; Hoggan, E. N.; Clark, N. A.; Rieker, T. P.; Walba, D. M.; Bowman, C. N. *Science* (in press).

(31) Guymon, C. A.; Hoggan, E. N.; Bowman, C. N. *Macromolecules* (submitted).

(32) Dahl, I.; Lagerwall, S. T.; Skarp, K. *Phys. Rev. A* **1987**, *36*, 4380.

(33) Clark, N. A.; Lagerwall, S. T. *Appl. Phys. Lett.* **1980**, *36*, 899.

(34) Hikmet, R. A. M.; Lub, J. *J. Appl. Phys.* **1995**, *77*, 6234.

(35) Guymon, C. A.; Dougan, L. A.; Hoggan, E. N.; Bowman, C. N. (in preparation).

Chapter 3

Solid-State NMR Spectroscopy for Characterization of Acrylate Reactions

J. Eric Dietz, Brett A. Cowans, Robert A. Scott, and Nicholas A. Peppas[1]

Polymer Science and Engineering Laboratories, School of Chemical Engineering, Purdue University, West Lafayette, IN 47907–1283

We have examined the polymerization of certain multi(meth)acrylates using solid state ^{13}C NMR spectroscopy. A quantitative single pulse method with gated decoupling was used to determine the number of unreacted double bonds in the polymer network. The heat of polymerization for the (meth)acrylate double bonds was verified using a combination of NMR spectroscopy and differential scanning photocalorimetry (DPC) experiments. The NMR method was also used to monitor the effects of aging or thermal cycling in these networks.

Crosslinked polymer networks formed from multifunctional acrylates are completely insoluble. Consequently, solid-state nuclear magnetic resonance (NMR) spectroscopy becomes an attractive method to determine the degree of crosslinking of such polymers (*1-4*). Solid-state NMR spectroscopy has been used to study the homopolymerization kinetics of various diacrylates and to distinguish between constrained and unconstrained, or unreacted double bonds in polymers (*5,6*). Solid-state NMR techniques can also be used to determine the domain sizes of different polymer phases and to determine the presence of microgels within a polymultiacrylate sample (*7*). The results of solid-state NMR experiments have also been correlated to dynamic mechanical analysis measurements of the glass transition (*1,8,9*) of various polydiacrylates.

In the present work, we use quantitative solid-state ^{13}C NMR spectroscopy to study the polymerization process of multiacrylates and the effects of thermal history/aging on the free radical life in polymultiacrylates.

Experimental

1,1,1-Trimethylolpropane triacrylate (TrMPTrA, mw 296.3, Polysciences Inc., Warrington, PA) was mixed with 1 wt% 2,2-dimethoxy-2-phenyl-acetophenone (DMPA, mw 256.3, Aldrich Chemical Co., Milwaukee, WI) as a photoinitiator. For the photocalorimetry experiments, a quantity of approximately 2 µl of this mixture was transferred to a differential photocalorimeter pan (DPC, model 930, TA Instruments, Wilmington, DE), and an empty pan was used as a reference. The light intensity was

---

[1]Corresponding author

adjusted to 1 mW/cm^2. The monomer and photoinitiator were stabilized for 15 minutes in a nitrogen atmosphere inside the DPC sample chamber and then the shutter was opened and the photopolymerization reaction was followed by determining the heat flow as a function of time. Graphs of the rate versus time were integrated to generate the conversion versus time as described before (*10*).

For NMR spectroscopic experiments, a thin film of pTrMPTrA was prepared by reacting a quantity of monomer and photoinitiator confined between glass plates with ~ 1 mm separation. The polymerization conditions were the same as those for the photocalorimetry experiments. After 1 hour of UV exposure, the film was removed from the plates and ground to a fine powder using a mortar and pestle. A solid-state ^{13}C NMR spectrum of the powder was obtained immediately, as described below. The remaining polymer powder was divided into two portions, one of which was stored under atmospheric conditions. The other portion was stored under N$_2$. After one week, ^{13}C spectra were again obtained for each of these polymer samples. Both samples were then heated to 280 °C in a vacuum oven and analyzed once more by ^{13}C NMR spectroscopy.

Solid-state ^{13}C NMR spectra were obtained on a General Electric Omega PSG 400 spectrometer operating at 100.6 MHz for ^{13}C nuclei using a Doty Scientific 5 mm high speed MAS probe. The various NMR experiments performed in this study are shown schematically in Figure 1. The carbon chemical shift scale (*11*) and 90 degree pulse width were calibrated using adamantane. The 90 degree pulse width was 5.0 μs. The pulse delay and the acquisition time for the single pulse experiment depicted in Figure 1A were chosen so as to minimize effects due to varying ^{13}C relaxation times and to differential NOE effects. In the absence of such effects, single pulse experiments can provide quantitative relative concentration information based on peak areas (*12*). A long pulse delay permits complete relaxation of the various ^{13}C nuclei. A short acquisition time with gated ^{1}H decoupling, in addition to the long pulse delay, provides decoupled spectra without differential nuclear Overhauser enhancements affecting the relative carbon intensities.

In order to determine the appropriate pulse delay, ^{1}H and ^{13}C spin lattice (T$_1$) relaxation times were measured by cross polarization methods (*13*), as shown in Figures 1b and 1c. The T$_1$ data was acquired with a cross polarization field strength of 50 kHz, a contact time of 1 ms, and a 5.5 kHz spin rate. All proton T$_1$ values were on the order of 1 s, while carbon T$_1$'s ranged from 10 to 30 s. The conjugated and unconjugated carbonyl groups had relaxation times of 13 s and 29 s, respectively. Based on these values, spectra were initially acquired with a pulse delay of 300 s. However, subsequent experiments performed with a pulse delay of 100 s gave spectra and curve fit areas identical to those obtained with the longer pulse delay. Therefore, in order to reduce the experiment time to a reasonable value, the 100 s pulse delay was used.

Results and Discussion

Spectra of the TrMPTrA monomer and polymer network are shown in Figure 2. Both spectra were acquired in the Doty MAS probe using the single pulse experiment in Figure 1a. The monomer spectrum was acquired non-spinning with 128 scans and a 10 s pulse delay. The liquid monomer shows correspondingly sharp resonances at 166 (CO$_2$), 131 (CH$_2$), 129 (CH), 64 (OCH$_2$), 41 (C), 24 (CH$_2$) and 8 (CH$_3$) ppm. The polymer sample was prepared as a thin film and ground into a powder. The polymer spectrum in Figure 2 was acquired spinning at 5.5 kHz with 960 scans and a 300 s pulse delay. In this spectrum, the broad resonance at 174 ppm is assigned to those carbonyl carbons adjacent to reacted double bonds, i.e. unconjugated carbonyl

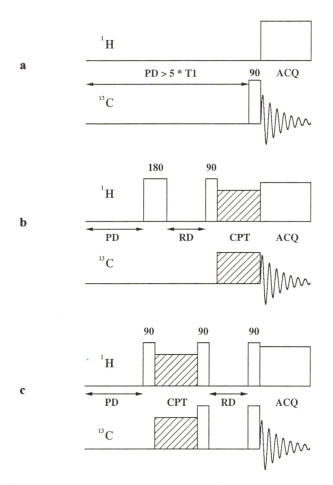

Figure 1. Schematic diagram of the solid-state NMR pulse sequences for (a) quantitative single pulse ^{13}C observe with gated ^{1}H decoupling and (b) ^{1}H T_1 and (c) ^{13}C T_1 determinations via cross polarization.

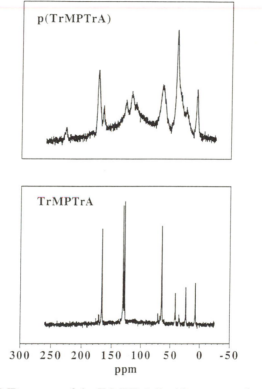

Figure 2. ^{13}C NMR spectra of the TrMPTrA liquid monomer (lower) and solid polymer (upper).

carbons. The resonance for those carbonyl groups which remain conjugated also becomes broader but their chemical shift remains constant at 166 ppm. Thus, the polymerization reaction can easily be monitored by comparing the relative intensities of the resonances. The only observable spinning side bands in this spectrum arise from these carbonyl carbons and appear at 228 and 120 ppm. The olefinic carbons at 131 and 129 ppm also decrease in intensity as broad, multiple methylene resonances appear in the region between 30 and 50 ppm. However, these resonances are not well resolved and are therefore less useful for monitoring the reaction. The extremely broad resonance observed at about 115 ppm in the polymer spectrum is do to the Kel-F housing in the probe.

Solid-state ^{13}C NMR spectroscopy was used to study the conversion of the monomer during polymerization, and the results were correlated with DPC results. The monomer was first reacted in the DPC, allowing for measurement of the heat of polymerization. Several DPC samples were combined to give approximately 25 mg of sample for the NMR experiments. A single pulse NMR spectrum was acquired with 1768 scans and a 100 s pulse delay. The double bond conversion for the DPC samples was determined from the relative ratio of reacted and unreacted carbonyl groups, which was calculated from a nonlinear, least squares, gaussian curve fit through the NMR data (14). The conversion as measured by NMR was found to be approximately 50%, while the average enthalpy of reaction determined by DPC was 413 J/g. These results gave a calculated reaction enthalpy for TrMPTrA of 826 J/g. This value compares favorably with the value of 873 J/g, which was determined by the technique proposed by Moore (15). Therefore, we have an independent method of determining the heat of reaction for this photopolymerization process.

Solid-state NMR spectroscopy was also used to examine the post reaction behavior of pTrMPTrA samples prepared in bulk as thin films, as described in the experimental. All of the spectra in this aging study required a minimum of 720 scans on approximately 50 mg of sample with a 100 s pulse delay to achieve adequate signal/noise. Under these conditions, reliable peak areas could be obtained from the curve fits of the carbonyl region. Figure 3 depicts the evolution of the solid state spectrum of the sample stored under N_2 over time and upon heating. The area of the peak at 174 ppm for the carbonyl adjacent to the reacted double bond increases as the peak at 166 ppm for pendant unsaturation decreases. The results of the aging study are given in Table I.

Table I. Double bond conversion as measured by solid state NMR spectroscopy for pTrMPTrA samples with various storage and treatment conditions

Polymer Thermal History	Percent Double Bond Conversion for Various Polymer Storage/Treatment Conditions	
	Air	N_2
2.5 h after reaction	73	-
1 week after reaction	75	82
Thermal treatment 1 week after reaction (280 °C)	93	92

The double bond conversion as measured by NMR immediately following the reaction was 73%. The conversion is significantly higher than that measured in the DPC experiments described above due to the more substantial rise in temperature accompanying polymerization on a larger scale.

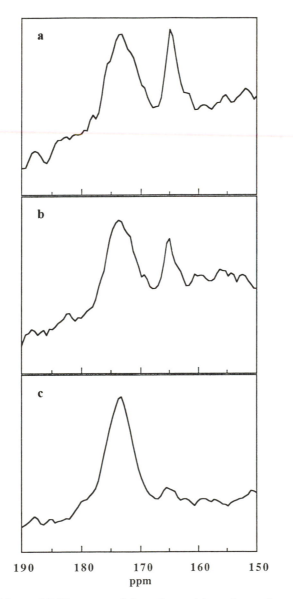

Figure 3. Solid-state NMR spectra of the polymer (a) one hour after reaction, (b) one week after reaction (N_2 storage), and (c) one week following reaction and thermal treatment at 5 °C/min to 280 °C (N_2 storage).

NMR analysis of the polymer sample subsequently stored in air for one week at room temperature gave a double bond conversion of 75%, within experimental error of the value obtained immediately after polymerization. These values were shown to be reproducible. However, the sample stored under N_2 following polymerization showed an increase in double bond conversion to a value of 82% one week after polymerization. This result suggests that additional polymerization occurs over time in pTrMPTrA networks in the absence of initiation, due most likely to the presence of trapped radicals. These radicals are quickly scavenged in the presence of oxygen. After both polymer samples were subjected to the heating profile described in the experimental, the conversion of double bonds was 93% for the sample that had been stored under air and 92% for the sample that had been stored under N_2. The difference in measured double bond conversion between the two samples is within the experimental error of at least ± 2%. Clearly, additional polymerization occurs through residual unsaturation upon heating to elevated temperatures.

We have found that combined solid-state NMR spectroscopy and DPC results can be used to calculate reaction enthalpies which are in close agreement with Moore (15). Furthermore, we find that the degree of conversion can be significantly affected by post reaction thermal processing. We conclude that the single pulse solid-state NMR spectroscopy can provide reliable, reproducible, and quantitative information about these highly crosslinked, insoluble, polymer networks.

Acknowledgments

This work was supported by grant No. CTS 93-11563 from the National Science Foundation.

Literature Cited

1.   Brar, A. S.; Sunita, J. *J. Polym. Sci.: Polym. Chem.* **1992**, *20*, 2549.
2.   Lungu, A.; Neckers, D. C. *Macromolecules* **1995**, *28*, 8147.
3.   Lungu, A.; Neckers, D. C. *J. Coatings Tech.* **1995**, *67*, 29.
4.   Yokota, K.; Abe, A.; Hosaka, S.; Sakai, I.; Saito, H. *Macromolecules* **1978**, *11*, 95 .
5.   Simon, G. P.; Allen, P. E. M.; Bennett, D. J.; Williams, D. R. G.; Williams, E. H. *Macromolecules* **1989**, *22*, 3555.
6.   Allen, P. E. M.; Bennett, D. J.; Hagias, S.; Hounslow, A. M.; Ross, G. S.; Simon, G. P.; Williams, D. R. G.; Williams, E. H. *Eur. Polym. J.* **1989**, *25*, 785.
7.   Ehrmannm, M.; Galin, J. -C.; Meurer, B. *Macromolecules* **1993**, *26*, 988 .
8.   Harrell, J. W.; Choudbury, M.; Ahuja, S.; Walter, W. *J. Polym. Sci., Polym. Phys.* **1991**, *29*, 1039.
9.   Shi, J. F.; Dickinson, L. C.; MacKnight, W. J.; Zang, C.; Liu, Y.; Chin, Y. H.; Jones, A. A.; Inglefield, P. T. *Macromolecules* **1993**, *26*, 1008.
10.   Kurdikar, D.; Peppas, N. A.; *Macromolecules* **1994**, *27*, 733.
11.   Earl, W. L.; VanderHart, D. L. *J. Magn. Reson.* **1982**, *48*, 35.
12.   Traficante, D. *Concepts in Magnetic Resonance* **1992**, *4*, 153.
13.   Torchia, D. H. *J. Magn. Reson.* **1978**, *30*, 613.
14.   MacDonald, J. C. *J. Magn. Reson.* **1980**, *38*, 381.
15.   Moore, J. E. In *Chemistry and Properties of Crosslinked Polymers*; Editor, S.S. Labana; Academic Press: New York, 1977; 535.

Chapter 4

The Effects of Initiator and Diluent on the Photopolymerization of 2-Hydroxyethyl Methacrylate and on Properties of Hydrogels Obtained

Yu-Chin Lai and Edmond T. Quinn

Contact Lens Division, Bausch & Lomb, Inc., Rochester, NY 14692–0450

A variety of photo-initiators and water-soluble diluents were used to study the photopolymerization of 2-hydroxyethyl methacrylate (HEMA). The rate of polymerization can be correlated to the structures as well as the concentration of initiators and diluents. It was found that, in terms of rate of polymerization, 2,2-dimethoxy-2-phenyl acetophenone and glycerine were the best initiator and the best diluent among all initiators and diluents evaluated. The rate of polymerization was found to have a great impact on mechanical properties, while it showed only minor effects on water contents of hydrogels derived from HEMA.

2-Hydroxyethyl methacrylate, (HEMA, **1**), is the most

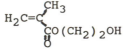

1

widely used UV-polymerizable hydrophilic monomer in making hydrogel lenses (*1*). However, the kinetics of its polymerization was not well studied. The effect of UV initiator and its concentration; and the effect of diluent and its concentration, on the rate of polymerization and on properties of hydrogels derived, are not well understood. In this paper, these aspects were addressed. Initiators and diluents which led to faster photo-polymerization of HEMA were also assessed.

Experimental

Monomers. HEMA was purified by aluminum oxide chromatography, followed by distillation under reduced pressure. All UV initiators and organic diluents were used as received.

Monomer Mixes for Polymerization Studies. A mixture containing HEMA and glycerine at a weight ratio of 85/15, and ethylene glycol dimethacrylate (EGDMA) at 0.4 %, and an initiator at 0.2 % of the weight of HEMA was the standard monomer mix used in photo-polymerization studies. Other variations were made by varying the initiator or the diluent concentration while keeping other components at the same relative concentrations, unless otherwise specified.

Polymerization Kinetics- by Photo Differential Scanning Calorimetry (Photo-DSC). In a DuPont differential scanning calorimetry unit, a monomer mix of 40 ul (40 mg) was placed in a sample pan at ambient temperature. Under nitrogen atmosphere, it was polymerized under UV, using a Sylvania F4T5 lamp (which gives a broad emission between 310-400 nm with peak at 350 nm). The effective light intensity was measured as 410 microwatts under this setup. The exothermic profile was monitored. The time when maximum polymerization rate (peak time) and when the polymerization stalled (stall time) were recorded. The % conversion was obtained by the integration of the exothermic profile. A typical exothermic profile for HEMA polymerization is shown in Figure 1 which also exemplified the definitions of peak time and stall time.

Preparation of Hydrogel Films Containing HEMA. Selected monomer mixes containing HEMA, EGDMA, glycerine and different initiators were prepared. The mixes were placed between two silane-treated glass plates and cured under UV (4000 microwatts) for 1 hour. The films were extracted with boiling water for 4 hours and then placed in a phosphate buffered saline (pH 7.3) before testing.

Characterization of Hydrogel Films. Mechanical testing was conducted in buffered saline on an Instron instrument, according to the modified ASTM D-1708 (tensile) and D-1938 (tear) and were reported in g/mm^2 for modulus and g/mm for tear strength. The water contents and the amount of extractables were measured gravimetrically.

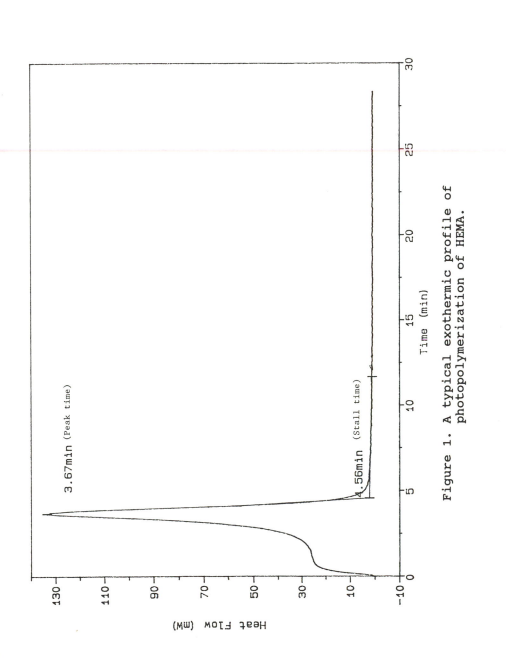

Figure 1. A typical exothermic profile of photopolymerization of HEMA.

Result and Discussion

The Effect of Initiators on the Polymerization of HEMA.
In terms of mechanism of radical generation, three
types of photo-initiator are frequently used in the
polymerization of vinyl monomers: PI_1 (α-cleavage,
unimolecular), PI_2 (H-abstraction, bimolecular), and
the radical cation type. A PI_1-type initiator is an
initiator which causes homolytic cleavage of a chemical
bond to form reactive radicals. PI_2-type initiators
usually are pairs of compounds containing a photosensitizer and an amine with an -hydrogen. The
sensitizer (usually is an aromatic ketone) forms an
exciplex with the amine after absorbing UV light, which
then cleaves to form an amine radical and initiates
polymerization. The radical cation initiators are
aryloxonium salts, such as triarylsulfonium
hexafluoroantimonate, which generate a radical cation
(usually is H^+) and initiates polymerization. Table I
lists the chemical structure, acronym, and absorption
maximum of initiators of three types used in this
study. Among them, benzoin methyl ether is used most
often for the polymerization of HEMA to give contact
lens. All initiators were screened first by Photo-DSC
using a standard HEMA monomer mix consists of
HEMA/Glycerine/EGDMA/photo-initiator at 85/15/0.34/0.17
(by weight). The UV lamp used (Sylvania F4T5 lamp) was
that with a broad emission spectrum between 310-400 nm.
The peak time in the exothermic profile, which is the
time when the rate of polymerization was fastest, was
recorded. Because peak time is indicative of the rate
of polymerization, initiators can be ranked by the peak
time in a photo-polymerization.

Table 2 summarizes the peak times for the photopolymerization of HEMA. Among initiators with
structures known, the ranking of more active
initiators, in terms of decreasing polymerization rate
was:

2,2-dimethoxy-2-phenyl acetophenone > benzoin methyl
ether $\geq$ 1-hydroxycyclohexane acetophenone = 2-hydroxy-
2,2-dimethyl acetophenone > TXN/MDEA

It is interesting to note that among initiators
studied, a benzyl ketal type initiator (2,2-dimethoxy-
2-phenyl acetophenone) is more active than other
initiators of acetophenone type. 2,2-Dimethoxy-2-
phenyl acetophenone is more active than BME, probably
due to the presence of the extra methoxy group on the -
carbon, which gives a more active phenyldimethoxy
methyl radical (and it further cleaves to give an even
more active methyl radical) than the phenylmethoxy
methyl radical from BME (4). It is known that the
phenylmethoxy methyl radical dimerizes easily and thus
loses some of its role as a radical in a

Table I. UV initiators used in the photopolymerization of HEMA

Initiator[2] (Acronym/Trade Name)	Structure	Abs Max (nm)
PI1 Type		
Benzoin Methyl Ether (BME)		323
2,2-dimethoxy-2-phenyl acetophenone (Irgacure-651)		335
2-hydroxy-2,2-dimethyl acetophenone (Darocur-1173)		320
1-hydroxycyclohexane acetophenone (Irgacure-184)		326
2,2-diethoxy acetophenone (DEAP)		323
4-(2-hydroxyehtoxy) phenyl-2-hydroxy-2-propyl ketone (Darocur-2959)		<313
Mixture of aromatic Ketones (Unknown structure)		
Darocur-1116		313
Darocur-1664		383
Darocur-2273		383
Darocur-3331		<313
PI2 Type		
Thioxanthne-9-one (TXN)		378
N-methyl diethanolamine (MDEA)		–
Radical Cation Type		
Triarylsulfonium hexafluorophosphate (UVI-6990, Cyracure)		–

Table II. Peak time of the Photo-DSC of HEMA monomer
mix with different of UV initiators. Formulation:
HEMA/Glycerine/EGDMA/initiator 85/15/0.34/x

Initiator*	Peak time (minute)			
Amount	.17	.34	.51	.68
Irgacure-651	2.89	2.28	2.58	2.58
BME	3.67	3.16	3.07	3.00
Irgacure-184	3.94	3.28	3.20	3.06
D-1173	4.01	3.36	3.27	3.20
DEAP			5.50	
D-1116	4.89			
D-1664	5.25			
D-2273	7.23			
D-2959	10.48			
D-3331	10.52	8.51	8.31	8.38
TXN/MDEA		5.24**		
UVI-6990		***		

* D means Darocur. Refer to Table 1 for
  chemical structures
** at 0.17 % each.
*** No meaningful peak time obtained

polymerization(4). The two hydroxyacetophenones
mentioned have very similar structure and thus showed
the same reactivity towards HEMA. From this study, it
was also found that initiators of the PI_2 and radical-
cation type did not polymerize HEMA as fast as some
very active initiators of the PI_1 type.

**The Effect of Initiator Concentration on the Rate of
Polymerization.** BME, 2,2-dimethoxy-2-phenyl
acetophenone, 2-hydroxy-2,2-dimethyl acetophenone and
Darocur-3331 (its structure is not known) were chosen
for further evaluation on the effect of initiator
concentration on the rate of polymerization. They were
chosen because they are the among the most active or
the least active initiator in polymerizing HEMA.

Table II also lists the peak times from these
studies. As expected, the rate of polymerization (by
peak time) increased as the concentration of initiator
was increased until screening effects started to appear
(between 0.34 to 0.68 %). In term of rate of
polymerization, the relative ranking ofinitiators
remained the same and 2,2-dimethoxy-2-phenyl
acetophenone was the most active.

Figure 2 further demonstrates the relationship
between rate of polymerization (peak times), and the
initiator concentration. Although trivial, it is
interesting to note that the polymerization rate
correlates well with the overlap between the emission

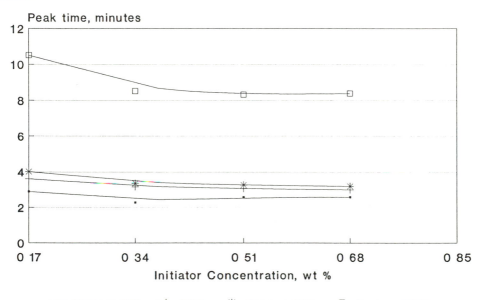

Figure 2. The effect of initiator on the peak time in the Photo-DSC for HEMA polymerization.
(see Table I for the chemical structures of initiators)

spectrum of the light source and the absorption
spectrum of a photo-initiator (concentration per 100 mL
of methanol), which were shown in Figure 3. The more
overlap of these two spectra (light source and UV
initiator), the faster the polymerization rate.

The Effect of Light Source on Curing Rate. As stated
earlier, Sylvania F4T5 was the lamp used in the
standard Photo-DSC measurements and this lamp had a
broad emission spectrum centered at 350 nm. When an
alternate lamp (GE F4T5) with an emission peak at 365
nm was used, the ranking of initiators, in terms of
peak time, remained the same. Table III lists
representative results from these experiments. The
emission spectrum of this GE lamp was also shown in
Figure 3.

**Table III. Peak time of the Photo-DSC of HEMA monomer
mix with different UV initiators using two different
lamps. Formulation: HEMA/Glycerine/EGDMA/initiator
85/15/0.34/0.17**

Initiator	Peak time (minutes)	
	Sylvania	GE
BME	3.67	5.00
D-1173	4.01	6.01
Irgacure-651	2.89	3.73
TXN/MDEA	5.24	6.21

The Effect of Diluents on the Curing of HEMA.
Generally, a non-reactive diluent is added to a monomer
to decrease the glass transitions of a growing polymer
during polymerization, thus helping driving the
polymerization to completion at a lower temperature.
The original HEMA formulation used in making hydrogel
lenses contained 15 % Glycerine. The choice of
glycerine at 15 % was based on manufacturing
requirements (having appripriate viscosity). It was
not known if glycerine at 15 % is at optimim in terms
of polymerization kinetics. In this study, water-
soluble diluents covering a reasonable range of
polarity, such as water, ethanol, ethylene glycol,
diethylene glycol, glycerine and N-methyl pyrrolidone
were used and compared. To make a comparison
meaningful, the standard monomer mix used was
HEMA/diluent/EGDMA/BME at 85/15/0.17/0.34. The wt % of
BME in the monomer mixes was maintained constant as the
diluent level was adjusted.
    Table IV lists the solubility parameters (δ) of
these diluents (3) and the peak times for polymerizing
HEMA monomer mixes with different diluents. The

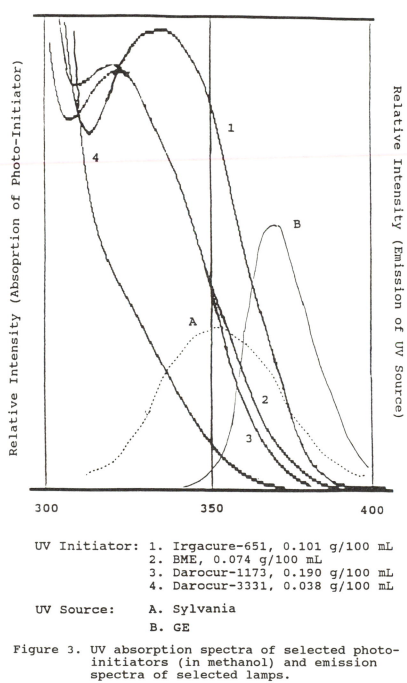

UV Initiator: 1. Irgacure-651, 0.101 g/100 mL
2. BME, 0.074 g/100 mL
3. Darocur-1173, 0.190 g/100 mL
4. Darocur-3331, 0.038 g/100 mL

UV Source: A. Sylvania
               B. GE

Figure 3. UV absorption spectra of selected photoinitiators (in methanol) and emission
spectra of selected lamps.

relationship between peak time and diluent levels was
further shown in Figure 4. The peak times of these gel
tests indicated that:
        a) Except for polymerizing HEMA in glycerine,
normal reaction kinetics was followed. When glycerine
was used, the rate of polymerization increased
initially increased, then leveled off (around 25-30 %
of glycerine).
.   b) In terms of rate of polymerization, glycerine
is the best diluent among those studied.
        c) Regardless of difference in polarity, ethylene
glycol, water, diethylene glycol and ethanol gave very
similar rate of polymerization rate for HEMA at all
diluent levels. The most polar diluent, N-methyl
pyrrolidone, gave the slowest curing rate of HEMA.

**Table IV. Peak times in the Photo-DSC of HEMA with
different diluents at 5, 15 and 25 %. Formulation
HEMA/EGDMA/BME/Diluent at 85/0.34/0.17**

Diluent	(δ)	Peak time (minutes)		
		5 %	15%	25 %
No			(4.36)	
Glycerine	16.5	4.12	3.92	3.47
diethylene glycol	12.7	4.27	4.69	5.69
ethylene glycol	14.6	4.30	5.27	6.35
ethanol	12.7	4.30	5.60	6.20
water	22.9	4.26	5.19	6.15
NMP	23.4	4.45	6.42	9.26

        To help understanding the peculiar kinetic
behavior for the polymerization in glycerine, the glass
transitions (T_g's) of cured mass were measured. Table
V gives T_g's of Poly(HEMA) after the photo-DSC tests.
Pure poly(HEMA) had a Tg of 104 $^{\circ}$C. When glycerine at
15 % was used, the poly(HEMA) had a T_g of 59 $^{\circ}$C, which
was substantial higher than room temperature. As more
glycerine was added, the Tg's decreased dramatically.
They were 39, 30 or 26 $^{\circ}$C when glycerine used was 25,
35 or 45 % respectively. Once the T_g of the growing
polymer dropped to ambient, the polymerization rate
would not increase by using more glycerine. However,
when other diluent was used at 15 %, T_g's of cured mass
were at or below room temperature.
        Regardless of high Tg, HEMA polymerized the
fastest in glycerine. This could be, at least
partially, attributed to a template effect as shown in
Figure 5. Glycerine, with at least one extra OH groups
compared to other diluents, has stronger intermolecular
hydrogen bonding, and thus more likely to align more

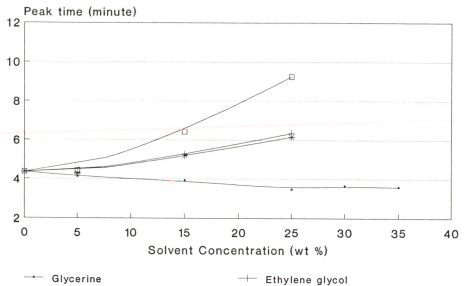

Figure 4. The effect of diluent on the peak time in
the Photo-DSC for HEMA polymerization.

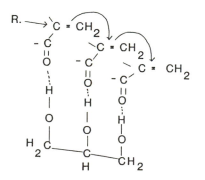

R. is the active radical
The methyl and ethylene gruops in HEMA
are not shown.

Figure 5. The proposed template effect of glycerine
on the polymerization of HEMA.

Table V.   Glass Transition temperatures
of cured HEMA samples. Formulation
HEMA/EGDMA/BME/Diluent : 85/0.34/0.17/X

Diluent	X	T_g (°C)
Glycerine	15	59
	25	39
	35	30
	45	26
Water	15	21
Ethylene glycol	15	15
NMP	15	29
(Dried polyHEMA)		104

HEMA molecules through hydrogen bonding with the
carbonyl group of HEMA. This may speed up the attack of
a double bond of HEMA moleculeby a neighboring HEMA
propagating radical. This also explains why higher
concentrations of glycerine speed up curing. Other OH-
containing diluents, such as ethylene glycol, are not
as polar, and not as effective in aligning HEMA
molecules.

**The Effect of Crosslinker Concentration on the Rate of
Polymerization.** Ethylene glycol dimethacrylate is used
most frequently as the crosslinker for HEMA
formulations useful in contact lens manufacturing. To
demonstrate the effect of crosslinker concentration on
the curing rate, formulations derived from
HEMA/Glycerine/BME at 85/15/0.17, while varying EGDMA
(from 0.34 to 0.68), the peak times were about the same
(3.73 and 3.61 minutes respectively). This is
reasonable due to the similarity in molecular structure
of the crosslinker and the monomer, and the low amount
of crosslinker used. The possible presence of other
crosslinker, such as the dimerization product of HEMA,
is even less a factor to be considered in
polymerization kinetics, due to low concentration
(normally much less than 0.1 %, in-house information).

**The % Conversion at Stall Time Vs the Rate of
Polymerization.** In a normal free radical
polymerization, the rate of polymerization stalls at
certain time when the mobility of molecules, including
those of propagating radicals, decrease to a certain
level. After that, the rate of polymerization
diminishes and it becomes a diffusion controlled
reaction. Table VI lists some representative % of
conversion at the stall time for randomly selected
HEMA-based monomer mixes with different initiators at
different concentration(s). They indicated that faster
polymerization rate led to higher conversion of

monomers at stall time. This phenomenon could be advantageous in driving the polymerization to completion particularly when the T_g of the polymerized mass rises sharply to higher than the ambient.

Table VI. Percentage conversion versus stall time in the Photo-DSC of HEMA-based monomer mixes. Formulation: HEMA/EGDMA/Glycerine at 85/0.34/15

Initiator/amount	Stall time (min)	% Conversion
Irgacure-651/0.51	3.40	96.5
Irgacure-651/0.68	1.99	98.0
BME/0.17	4.55	95.2
BME/0.51	4.30	95.8
DEAP/0.68	6.47	91.6
D-3331/0.68	10.1	90.2

The Effect of Rate of Polymerization on Properties of Hydrogels. Table VII gives properties of hydrogels based on the standard HEMA monomer mix, but cured with the 4 chosen initiators (the same 4 initiators as mentioned earlier) at concentrations of 0.17% to 0.51 %. The results indicated that the rate of polymerization gave only minor difference in water contents and extractables of the cured films. All hydrogel films were clear regardless of rate of polymerization. However, the rate of polymerization affected mechanical properties of hydrogels significantly. Faster polymerization (by using 2,2-dimethoxy-2-phenyl acetophenone at both concentration or by using BME at 0.51 %) gave hydrogels with much lower modulus and better tear strength, while slower polymerization gave films with higher modulus and lower tear strength. The changes in water content and mechanical properties may be due to decreases in

Table VII. The Effect of initiator and initiator concentration on properties of HEMA hydrogels: HEMA/Glycerine/EGDMA/Initiator at 85/15/0.34/X

Initiator	BME		Irgacure-651		D-1173		D-3331	
wt %	.17	.51	.17	.51	.17	.51	.17	.51
% Extract	1.4	1.7	1.8	2.2	0.6	1.4	1.0	1.5
% Water	39.7	40.3	39.6	40.2	38.5	38.1	38.7	36.0
Modulus	40	25	25	25	40	40	50	55
Tear	6	8	7	10	6	6	5	5

crosslinking density triggered by faster termination of
radicals containing crosslinker or by
cyclopolymerization of the crosslinker (EGDMA).

   The improved tear strength and lower modulus are
beneficial for some applications such as contact
lenses. Figure 6 illustrates a generalized
relationship between curing rate and mechanical
properties (both modulus and tear strength) of HEMA-
based hydrogels. The mechanical data were of those
HEMA-based hydrogels derived from monomer mixes using
different initiators and glycerine of varying
concentrations. The rates of polymerization were taken
from reciprocals of peak times in photo-DSC's. This
relationship clearly indicates that, for HEMA-based
monomer mixes which give hydrogels, the modulus of
hydrogel decreases and tear strength of hydrogel
increases as the rate of polymerization is increased.
It is interesting to note both modulus and tear
strength change fastest at the same polymerization rate
of HEMA (about 3.2 minutes in the peak time). However,
both modulus and tear strength leveled off at higher
rates of polymerization. This mechanical property -
rate of polymerization relationship probably is true
for most hydrogels. This relationship also indicates
the importance of polymerization rate in controlling
properties of a hydrogel. If it is desirable to
maintain the mechanical properties of a hydrogel, the
rate of polymerization still can be increased, while
offsetting the change in mechanical properties by
adjusting the level of crosslinkers. Table VIII lists
some representative HEMA-based hydrogel lenses
manufactured with monomer mixes containing either BME
or 2,2-dimethoxy-2-phenyl acetophenone as the initiator
at different levels. While formulation containing 2,2-
dimethoxy-2-phenyl acetophenone (formulation B) gave

Table VIII. HEMA hydrogel lenses fabricated from
monomer mixes containing different initiators.

Formulation	A	B	C
HEMA	85	85	85
Glycerine	15	15	15
EGDMA	0.34	0.34	0.68
BME	0.17	-	-
Irgacure-651*	-	0.68	0.68
Hydrogel Properties:			
Modulus	64	42	65
Tear strength	5.0	7.1	5.4

   * 2,2-dimethoxy-2-phenyl acetophenone

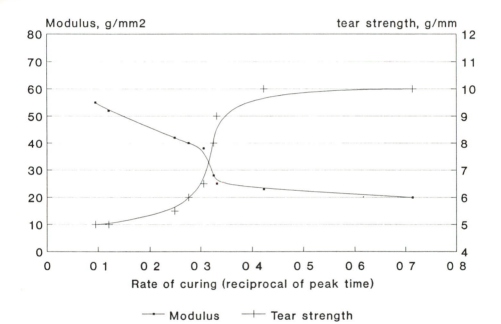

Figure 6. The relationship between modulus/tear
strength and the rate of polymerization.

hydrogel with modulus lower than that derived from
formulation A (using BME) when the same level of EGDMA
was used. The modulus for formulation with 2,2-
dimethoxy-2-phenyl acetophenone as the initiator could
be adjusted back that of formulation A when EGDMA was
doubled (Formulation C)
Conclusion.

For the photo-polymerization of HEMA, the use of more
active initiator and at higher concentration lead to
faster polymerization. WHen lamps which give emission
peaks around 350-365 nm, 2,2-dimethoxy-2-phenyl
acetophenone was found to be the best initiator in
terms of rate of polymerization. Glycerine as a
diluent gave the fastest rate of polymerization even
though it gave cured polymer with the highest Tg. A
template effect was suggested for this unusual kinetic
behavior. Fast photo-polymerization of HEMA led to a
reduction in modulus and increase in tear strength
while water content of hydrogel derived was maintained
constant.

Acknowledgments

The author acknowledged the help of the following
personnel: S. Hill for film and lens casting, C
Sevilla and M. Andrews for mechanical testing.

Literature Cited

1. Lai, Y.C., Wilson, A. C., and Zantos, S. Z. "*Contact
   Lens*" in Kirk Othmer Ency. Chem. Tech. 4th Edition.
   John Wiley & Sons, New York, NY. 1993, 7, 192-218.
2. a) Allen, N. "*Photoinitiators for Ultraviolet
   Curing*" in "Trends in Polymer Science", 1993, 1(7),
   206. b) Union Carbide product brochure.
3. Fox, K. L. J. Paint Tech., 1970, 42(541), 26.
4. Chang, C. H., Mar, A., Tiefenthaler, A., and
   Wostratzky, D. "Photoinitiators: Mechanisms and
   Applications", a publication form Ciba-Geigy
   Corporation, Hawthorne, New York.

Chapter 5

Reaction Behavior and Kinetic Modeling Studies of "Living" Radical Photopolymerizations

Anandkumar R. Kannurpatti, Michael D. Goodner, Hyun R. Lee, and Christopher N. Bowman[1]

Department of Chemical Engineering, University of Colorado, Campus Box 424, Boulder, CO 80309–0424

Free radical polymerizations of 2-hydroxyethyl methacrylate (HEMA) and diethylene glycol dimethacrylate (DEGDMA) initiated by a combination of a conventional initiator, 2,2-dimethoxy-2-phenyl acetophenone (DMPA), and N,N,N',N'- tetraethylthiuram disulfide (TED) were studied. TED generates dithiocarbamyl (DTC) radicals upon radiation which terminate with propagating carbon radicals to produce carbon-DTC groups (which can be reinitiated). The kinetics of these photopolymerizations were monitored by differential scanning calorimetry. It was observed that as the TED concentration was increased, the characteristic autoacceleration effect in both polymerizations was suppressed. The presence of the carbon-DTC radical termination pathway keeps the carbon radical concentration low, even when the carbon-carbon termination reaction becomes diffusion limited. As a result, the autoacceleration peak is suppressed in these polymerizations. Further, to understand the behavior of the kinetic constants during the polymerizations, k_p (propagation kinetic constant) and k_t (termination kinetic constant for carbon-carbon radical termination) were measured by after-effect experiments in samples initiated only by DMPA. Using these measured values of the kinetic constants in a free volume based model, the reaction behavior of these polymerizations was simulated to glean more information regarding the polymerization mechanism and the effects of the DTC radical concentration.

"Living" radical polymerizations have received considerable attention because they provide a convenient alternative for synthesizing block copolymers, polymers of narrow polydispersity and complex polymer structures (1-5). Because of their ability to initiate living free radical polymerizations, iniferters have been examined extensively after Otsu *et al.* (6) introduced them in 1982. In particular, dithiocarbamate derivatives have been studied more closely by several researchers. Lambrinos *et al.* (7) have examined the molecular weight evolution during the polymerization of *n*-butyl acrylate using *p*-xylylene bis(*N,N*-diethyl dithiocarbamate). Further, it has also been noted that during chain extensions performed over hours using polymeric iniferters (dithiocarbamate derivatives) side reactions that produce carbon disulfide have been observed (8,9). However, during

[1]Corresponding author

homopolymerizations of butyl acrylate using *n*-butyl (*N,N*-diethyl dithiocarbamyl) propionate Dika Manga *et al.* (9) observed negligible trace amounts of carbon disulfide. They also noted that to obtain polymers of high functionalities, the polymerization time should be low and the monomer concentration should be high.

While there have been several studies on the synthesis of block copolymers and on the molecular weight evolution during solution as well as bulk polymerizations (initiated by iniferters), there have been only a few studies of the rate behavior and kinetic parameters of bulk polymerizations initiated by iniferters. In this paper, the kinetics and rate behavior of a two-component initiation system that produces an *in situ* "living" radical polymerization are discussed. Also, a model that incorporates the effect of diffusion limitations on the kinetic constants is proposed and used to enhance understanding of the living radical polymerization mechanism.

A typical photoreactive iniferter such as *p*-xylylene bis(*N,N*-diethyl dithiocarbamate) (XDT) absorbs ultraviolet light and generates a reactive carbon radical and a less reactive dithiocarbamyl (DTC) radical. As the carbon radical reacts through the double bonds of the monomer, the DTC radical acts as a radical scavenger and keeps the radical concentration and the rate of polymerization low. The carbon-DTC radical termination produces reinitiatable (by UV light) polymer chains with DTC end-groups throughout the reaction. However, bimolecular carbon-carbon radical termination results in a polymer that cannot reinitiate. This reaction leads to the deactivation of the iniferter and prevents the polymerization from being "living". In a recent study (10), it was found that as the concentration of DTC radicals was increased by adding tetraethylthiuram disulfide (TED generates only DTC radicals on photolysis) to XDT the "living" nature of the polymerization 2-hydroxyethyl methacrylate (HEMA) was improved. Also, it was found that if there were an excess of DTC radicals in the reacting system, the carbon-DTC radical termination dominated over carbon-carbon radical termination. Similar observations were reported by Doi *et al.* (11) in their study of methyl acrylate polymerizations using a combination of benzyl *N,N*-diethyl dithiocarbamate and TED.

In this work, the kinetics of these reactions are closely examined by monitoring photopolymerizations initiated by a two-component system consisting of a conventional photoinitiator, such as 2,2-dimethoxy-2-phenyl acetophenone (DMPA) and TED. By examining the polymerization kinetics in detail, further understanding of the complex initiation and termination reactions can be achieved. The monomers discussed in this manuscript are 2-hydroxyethyl methacrylate (HEMA), which forms a linear polymer upon polymerization, and diethylene glycol dimethacrylate (DEGDMA), which forms a crosslinked network upon polymerization.

By examining HEMA, useful information regarding the "living" radical polymerizations, which can influence their potential use in the synthesis of block copolymers, and polymers with low polydispersity, can be gleaned. Crosslinking "living" radical polymerizations are of interest as they can be used to synthesize polymers with no trapped radicals and thus, facilitate characterization of the evolution of structure as well as properties in these networks. To provide a better understanding of these crosslinking polymerizations, the studies on DEGDMA have been performed.

During conventional polymerizations of both HEMA and DEGDMA, complications resulting from diffusion limitations to termination and propagation are observed. Features such as autoacceleration, autodeceleration and incomplete conversion of double bonds characterize the rate behavior of these polymerizations. As TED is added to the reacting system, the carbon-DTC radical termination reaction is introduced. Diffusion limitations to carbon-DTC radical combination are lower than those to carbon-carbon radical termination as the DTC radical is smaller and much more mobile than a typical polymeric carbon radical. As a result, the cross-termination of carbon radicals with DTC radicals is preferred to bimolecular carbon-carbon radical termination. This shift in the termination mechanism results in the suppression of the autoacceleration effect and a decrease in the overall rate of polymerization. To understand further this mechanism, a model incorporating the diffusion limitations in the bulk polymerizing systems studied is presented.

Experimental

The monomers studied, 2-hydroxyethyl methacrylate (HEMA) and diethylene glycol dimethacrylate (DEGDMA), were obtained from Aldrich (Milwaukee, WI) and Polysciences, Inc. (Warrington, PA), respectively, and were used after dehibition to remove the hydroquinone inhibitor.   2,2-Dimethoxy-2-phenyl acetphenone (DMPA), the conventional initiator used in this study, was obtained from Ciba-Geigy (Hawthorne, NY) and the tetraethylthiuram disufide (TED) was obtained from Aldrich.

Kinetic experiments were performed in a differential photo-scanning calorimeter (DPSC). The DPSC is equipped with a monochromator which was used to obtain 365 nm light for the polymerizations. All polymerizations were performed at a light intensity of approximately 4 mW/cm^2. The rate of polymerization is monitored by following the rate of heat evolved in these highly exothermic reactions. The standard heat of reaction for methacrylates is -13.1 kcal/mol (12). DMPA concentrations of 0.5 wt% were used, with varying concentrations of TED. Samples ranging from 2 to 3 milligrams were used to ensure the validity of the thin film approximation. The samples in the DSC were purged prior to and throughout the reaction with nitrogen to prevent oxygen inhibition during the polymerization.

The rate of polymerization in the dark, *i.e.*, when the initiation is stopped, was used to uncouple the kinetic constants for termination and propagation. By performing these 'after-effect' experiments (13-15) at various stages in the reaction, the kinetic constants for bimolecular carbon-carbon termination and propagation were characterized as a function of conversion.

Model Development

The rate behavior is modeled using kinetic expressions based on elementary reactions of the species involved. Generation of radicals can occur through five different initiation mechanisms. First, species such as DMPA or TED can generate either two carbon radicals or two DTC radicals. If XDT-like initiators are considered, one carbon radical and one DTC radical are generated upon photolysis, and a similar reaction for reinitiation of DTC-terminated polymer chains exists. Lastly, initiation of polymer chains by DTC radicals should be included for completeness. These reactions can be summarized as:

$$I_{CC} \xrightarrow{\phi_{CC}} 2R_C \bullet \qquad\qquad (1)$$

$$I_{SS} \xrightarrow{\phi_{SS}} 2R_S \bullet \qquad\qquad (2)$$

$$I_{CS} \xrightarrow{\phi_{CS}} R_C \bullet + R_S \bullet \qquad\qquad (3)$$

$$R_S - P \xrightarrow{\phi_{SP}} R_C \bullet + R_S \bullet \qquad\qquad (4)$$

$$R_S \bullet + M \xrightarrow{k_{i,SM}} R_S - R_C \bullet \qquad\qquad (5)$$

$R_c\bullet$ and $R_s\bullet$ represent the carbon radicals and DTC radicals, respectively. I_{CC}, I_{CS} and I_{SS} are the possible initiator species (DMPA, XDT and TED, respectively), M represents the monomer, P the polymer and R_s-P the telechelic DTC-terminated polymer. ϕ_{XX} are the initator efficiencies of the various initiating (or telechelic) species and $k_{i,SM}$ is the kinetic constant for DTC monomer initiation. In this study, monomer initiation by DTC is assumed to be negligible (i.e., $k_{i,SM} = 0$). This assumption was verified by illuminating a mixture of HEMA and TED with ultraviolet light and noting that polymerization occuring on the time scale of these studies was negligible. Furthermore, all carbon radicals are considered to be equivalent, independent of length. Also, it is assumed that R_s-P behaves like I_{CS} with respect to initiation, i.e., they have the same initiator efficiency.

In the thin films used for these studies, the initiation rates are given by:

$$R_{i,XX} = 2\phi_{XX}\varepsilon_{XX}I_0 b[I_{XX}] \qquad (6)$$

where ε_{XX} are the molar absorptivities in l/(mol·cm), $I_0 b$ is the incident light intensity in moles of photons/(s·cm^2) and $[I_{XX}]$ is the initiator species concentration in mol/l.

Once the reaction of DTC radicals with monomer is neglected, propagation of radicals follows the mechanism seen in other radical photopolymerizations:

$$R_C \bullet + M \xrightarrow{\ k_p\ } R_C \bullet \qquad (7)$$

Termination of active radicals can occur via three mechanisms: carbon-carbon termination, cross termination and recombination of DTC radicals. These reactions appear as:

$$R_C \bullet + R_C \bullet \xrightarrow{\ k_{tCC}\ } P \qquad (8)$$

$$R_C \bullet + R_S \bullet \xrightarrow{\ k_{tCS}\ } R_S - P \qquad (9)$$

$$R_S \bullet + R_S \bullet \xrightarrow{\ k_{tSS}\ } I_{SS} \qquad (10)$$

In this study, molecular weight of the produced polymers will not be tracked over the course of the reaction. Thus, in order to simplify the model, chain transfer mechanisms will not be considered, along with side reactions, such as the production of carbon disulfide. Each of these reactions, as well as the molecular weight, plays a significant role in the iniferter polymerizations; however, to simplify the system, it is essential to examine only the core reactions which contribute significantly to the mechanism.

Because diffusion limitations exist, the kinetic constants for propagation and termination (all three modes) vary with free volume, and therefore conversion. To characterize the propagation and carbon-carbon termination kinetic constant behavior as conversion increases, after-effect experiments were performed. For typical photopolymerizations initiated only by DMPA, the time rate of change of polymerization rate when there is no initiation (after extinguishing the light source at some point in the reaction) gives k_t/k_p. For an initiating system (light still on) at the same conversion, rate versus monomer concentration gives $k_p/k_t^{1/2}$. Thus, the two values can be combined to yield k_p and k_t separately. This process can be repeated at several conversions to glean the general trend of kinetic constant evolution. Note that this technique determines only the carbon-carbon termination constant, as other modes modify the dark reaction kinetics.

In the model, the kinetic constants for propagation and termination are allowed to vary as a function of free volume, as suggested by Marten and Hamielec (16) and Anseth and Bowman (17). To account for diffusional limitations and still predict the non-diffusion controlled kinetics, the functional forms for the propagation and carbon-carbon termination kinetic constants are:

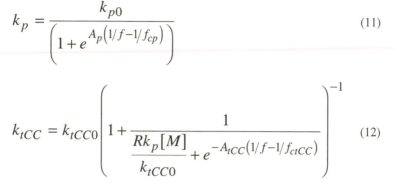

$$k_p = \frac{k_{p0}}{\left(1 + e^{A_p\left(1/f - 1/f_{cp}\right)}\right)} \tag{11}$$

$$k_{tCC} = k_{tCC0}\left(1 + \frac{1}{\dfrac{Rk_p[M]}{k_{tCC0}} + e^{-A_{tCC}\left(1/f - 1/f_{ctCC}\right)}}\right)^{-1} \tag{12}$$

In the expression determining k_p, the propagation kinetic constant, k_{p0} is the pre-exponential factor, A_p is a parameter determining the rate of decrease of k_p in the diffusion limited region, f is the fractional free volume of the system, and f_{cp} is the critical fractional free volume for propagation, the point at which diffusion limitations control propagation. The expression for k_{tCC} is similar, however a term is included to account for reaction diffusion, which is termination by the propagation of an active radical to another radical, rather than the diffusion of the macroradical. If R, the reaction diffusion parameter, is set to zero, the expression for k_{tCC} will reduce to the diffusional control form seen for k_p. The forms for carbon-DTC termination and DTC-DTC termination kinetics constants, k_{tCS} and k_{tSS}, have similar diffusion limitations as propagation, and thus will have the same form as k_p.

If R is known, it is possible to fit the parameters k_{p0}, k_{tCC0}, A_p, A_t, f_{cp} and f_{ct} using kinetic data from a single experiment. Thus, if the reaction diffusion parameter is known from the unsteady state after-effect experiments, the kinetic constant evolution can be determined as a function of free volume, and thus conversion. More details about this method will be published elsewhere (18).

Thus, the parameters characterizing propagation and carbon-carbon termination are experimentally determined, as are initiator molar absorptivities (from spectrophotometry). If k_{tSS0}, k_{tCS0}, A_{tSS}, A_{tCS}, f_{cSS} and f_{cCS} parameters for DTC-DTC and cross termination are intuitively chosen (based on comparisons with parameters for k_p and k_t) the system can be quantitatively modelled. The parameters used for the simulations presented here can be found in Table I.

Conservation equations are written for all reactive species: initiators, monomer, polymer carbon radicals and DTC radicals. They are integrated forward in time using the forward Euler technique, and the results can be presented as functions of either time or conversion. The results for these simulations are given in the following section.

Results and Discussion

When the polymerization is initiated by a conventional initiator such as DMPA, HEMA and DEGDMA display rate curves as shown in Figures 1 and 2. In these

Table I. Material and kinetic properties for HEMA, DEGDMA and the initiators used in this study

Material Properties for HEMA
$\rho_m = 1.073$ g/cm^3	$\rho_p = 1.15$ g/cm^3	$[M]_0 = 8.2$ mol/l
$T_{gm} = -60\ °C$	$T_{gp} = 55\ °C$	
$\alpha_m = 0.0005\ °C^{-1}$	$\alpha_p = 0.000075\ °C^{-1}$	

Material Properties for DEGDMA
$\rho_m = 1.061$ g/cm^3	$\rho_p = 1.32$ g/cm^3	$[M]_0 = 4.35$ mol/l
$T_{gm} = -100\ °C$	$T_{gp} = 225\ °C$	
$\alpha_m = 0.0005\ °C^{-1}$	$\alpha_p = 0.000075\ °C^{-1}$	

Kinetic Parameters for HEMA
$R = 4$
$k_{p0} = 1000$ l/mol·s	$A_p = 0.66$	$f_{cp} = 0.042$
$k_{tCC0} = 1.1 \times 10^6$ l/mol·s	$A_{tCC} = 1.2$	$f_{ctCC} = 0.060$
$k_{tCS0} = 1.0 \times 10^3$ l/mol·s	$A_{tCS} = 0.6$	$f_{ctCS} = 0.042$

Kinetic Parameters for DEGDMA
$R = 2$
$k_{p0} = 100$ l/mol·s	$A_p = 1.2$	$f_{cp} = 0.062$
$k_{tCC0} = 2.0 \times 10^4$ l/mol·s	$A_{tCC} = 2.7$	$f_{ctCC} = 0.089$

Initiator Properties
DMPA:	$\varepsilon = 150$ l/mol·cm	$\phi_{CC} = 0.6$
TED:	$\varepsilon = 120$ l/mol·cm	$\phi_{CS} = 1.0$
XDT:	$\varepsilon = 200$ l/mol·cm	$\phi_{SS} = 1.0$
R_S-P:	$\varepsilon = 100$ l/mol·cm	$\phi_{SP} = 1.0$

polymerizations, DMPA concentrations of 0.5 wt% and a light intensities of 4.2±0.2 mW/cm^2 were used. In the HEMA polymerization, the initiated carbon radicals propagate through the monomeric double bonds. As the polymerization proceeds, more polymer is formed, resulting in an increase in the viscosity of the system. At about 15% conversion, the carbon-carbon radical termination starts to become diffusion limited as the viscosity of the system steadily increases beyond a certain critical viscosity. Such diffusion limitations to carbon-carbon radical termination lead to a dramatic increase in the radical concentration. As a result, the rate of polymerization increases rapidly, as the rate is proportional to the radical concentration. This autoacceleration effect is observed in Figure 1 for HEMA polymerization. As the polymerization proceeds further, the system viscosity limits propagation, as monomer cannot diffuse to the relatively immobile radicals. The rate of polymerization decreases (as it is proportional to the propagation kinetic constant (k_p)) and continues to do so until no further reaction takes place because the system has vitrified.

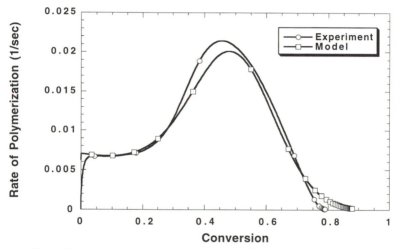

Figure 1. Experimental and simulated normalized rates of polymerization for HEMA initiated by DMPA.

The model captures the features seen in the experiment very well. The initial rate of polymerization and the conversion at the onset of autoacceleration are nearly identical with the experimentally generated values as is the rate of autodeceleration. The conversion and value of maximum rate are within 5%. The difference in maximum rate can be ascribed to the high sensitivity of autoacceleration on the A_t parameter; a small change in A_t can greatly influence the rate of polymerization during autoacceleration. The conversion at which the maximum rate occurs is dependent upon f_{cp}, *i.e.*, the free volume (and conversion) at which autodeceleration sets in. The simulated rate also shows a tail around 80% conversion, which can be ascribed to the DSC not capturing polymerization at high conversion and low rate.

Similar trends are observed in the case of DEGDMA (Figure 2); however, the autoacceleration peak begins very early in the reaction (less than 2% conversion) and to a greater extent than HEMA. Also, the autodeceleration takes place at lower conversions and final conversions as low as 60% may be observed. These differences are caused by the higher functionality of DEGDMA. The two methacrylate units in DEGDMA facilitate formation of highly crosslinked networks upon polymerization, while HEMA is difunctional and forms a linear polymer. Formation of highly crosslinked networks such as poly(DEGDMA) is associated with very early onset of autoacceleration as observed because the termination process becomes diffusion limited at very low conversion of double bonds. A complex heterogeneous structure evolves as polymerization proceeds and propagation becomes diffusion controlled, thus, leading to autodeceleration. Further, studies (19,20) have shown that in these polymerizations the k_{tCC} becomes proportional to the product of k_p and monomer concentration. In this regime, the termination becomes "reaction-diffusion" controlled, *i.e.*, the radicals are more mobile via propagation through double bonds than they are by any diffusive mechanism.

As in the case of HEMA polymerizations, the model predicts the experimentally observed rate successfully. The rates of autoacceleration and autodeceleration are predicted well by the model, and the initial rate and maximum rate are within 5%. A more pronounced 'shoulder' is seen in the model predictions than in the experiments. This can be attributed to DSC startup, going from zero rate to a high rate as the photoaccessory is turned on.

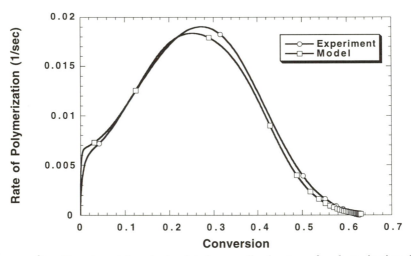

Figure 2. Experimental and simulated normalized rates of polymerization for DEGDMA initiated by DMPA.

The behavior of the termination kinetic constant (k_{tCC}) and propagation kinetic constant (k_p) are depicted in Figure 3 for HEMA polymerizations initiated only by DMPA. The kinetic constants were measured by monitoring the rate of polymerization in the dark and uncoupling the k_{tCC} and the k_p (13-15). It can be seen that the k_{tCC} starts decreasing at about 15% conversion and continues to drop throughout the reaction. The k_p on the other hand, remains roughly constant and decreases at higher conversions (beyond 50%). These effects are as expected because the diffusion limitations for termination of the kinetic chains become significant at lower conversions while the mass transfer limitations for the propagation of smaller monomer units are not significant until later in the reaction.

Having characterized the regular polymerization behavior of the monomers, experiments were performed to examine the effect of addition of TED to the polymerizing system. Figure 4 depicts rate curves for HEMA polymerizations with a

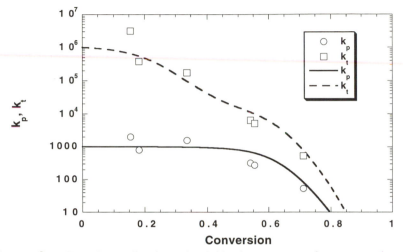

Figure 3. Experimentally determined kinetic constants for propagation and termination in DMPA-initiated HEMA polymerization.

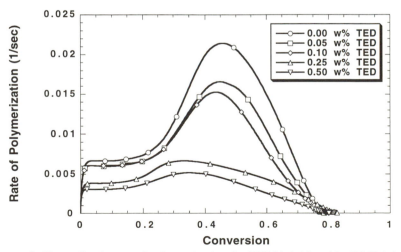

Figure 4. Normalized rates of polymerization for HEMA initiated by DMPA in the presence of increasing amounts of TED.

constant 0.5 wt% DMPA and increasing concentrations of TED. The TED concentration was varied from 0.05 to 0.5 wt%.

It can be observed that the initial rate of polymerization decreases and the autoacceleration peak is suppressed as the TED concentration is increased. The TED molecules generate dithiocarbamyl (DTC) radicals upon initiation. As a result, termination may occur by carbon-carbon combination which leads to a dead polymer and by carbon-DTC radical reaction which produces a reinitiatable ("living") polymer. The cross-termination of carbon-DTC radicals occurs early in the reaction (with the carbon-carbon radical termination), and this feature is observed by the suppression of the initial rate of polymerization. As the conversion increases, the viscosity of the system poses mass transfer limitations to the bimolecular termination of carbon radicals. As has been observed in Figure 3, this effect results in a decrease in the k_{tCC}. However, as the DTC radicals are small and mobile, the cross-termination does not become diffusion limited, $i.e.$, the kinetic constant for termination of carbon-DTC radicals, k_{tCS}, does not decrease. Therefore, the cross-termination becomes the dominant reaction pathway. This leads to a suppression of the autoacceleration peak as the carbon-DTC radical termination limits the carbon radical concentration to a low value, thus limiting the rate of polymerization. This observation is in accordance with results of previous studies (10) with XDT and TED, where it was found that when there was an excess of DTC radicals, the carbon radical concentration was lower and the cross-termination reaction was the dominant termination pathway.

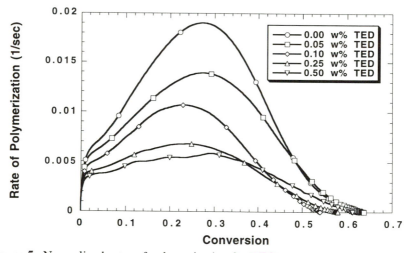

Figure 5. Normalized rates of polymerization for DEGDMA initiated by DMPA in the presence of increasing amounts of TED.

Figure 5 shows similar experimental rate data for the DEGDMA/DMPA/TED polymerization. As seen in the case of HEMA, TED addition decreases both the initial rate and the maximum rate of polymerization of DEGDMA. As described earlier, polymerization of DEGDMA results in a highly crosslinked polymer. The autoacceleration effect is characterisitc of highly crosslinked systems as the diffusional limitations reduce the carbon-carbon radical termination kinetic constant

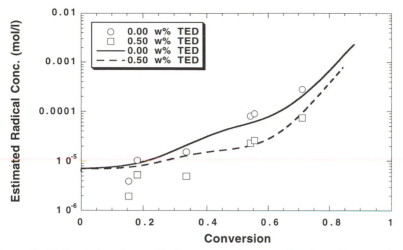

Figure 6. Estimated carbon radical concnetrations as a function of conversion for DMPA-initiated HEMA polymerization with and without TED present.

even in the very early stages of the polymerization. However, as the TED concentration is increased from 0.05 to 0.5 wt% it can seen that the autoacceleration effect is suppressed. At higher TED concentrations the autoacceleration effect is almost completely eliminated. These observations add further evidence to the dominance of the cross-termination mechanism in diffusion limited regimes.

Using the kinetic parameters measured by after-effect experiments, the radical concentration during the polymerization of HEMA can be estimated. As the propagation kinetic constant, k_p, does not change as the initiation system is changed, the propagating radical concentration can be calculated by evaluating the ratio of the rate of polymerization to the product $k_p[M]$. Here, $[M]$ is the monomer concentration which is dependent on the conversion during the reaction. Figure 6 compares estimated radical concentrations during HEMA polymerizations initiated by DMPA and by a combination of DMPA and TED. It can be observed that the presence of the TED suppresses the carbon radical concentration as expected. The DTC radicals terminate with the carbon radicals keeping the carbon radical concentration lower than it would be in regular polymerization with DMPA. This effect is observed from the beginning of the reaction, suggesting that the cross-termination mechanism is preferred very early in the polymerization.

Conclusions

In this paper, the kinetics and polymerization behavior of HEMA and DEGDMA initiated by a combination of DMPA (a conventional initiator) and TED (which produces DTC radicals) have been experimentally studied. Further, a free volume based kinetic model that incorporates diffusion limitations to propagation, termination by carbon-carbon radical combination and termination by carbon-DTC radical reaction has been developed to describe the polymerization behavior in these systems. In the model, all kinetic parameters except those for the carbon-DTC radical termination were experimentally determined. The agreement between the experiment and the model is very good.

From these experimental and modeling studies, the mechanism of the "living" free radical polymerizations initiated by a combination of TED and DMPA have been elucidated. The TED produces DTC radicals that preferentially cross-terminate with the propagating carbon radicals. By this cross-termination reaction, the carbon radical concentration is kept low (as was shown in figure 6) and the rate of polymerization is decreased, as is the autoacceleration effect. This suppression of the autoacceleration peak in HEMA polymerizations and, interestingly, in DEGDMA polymerization has been observed to increase as the TED concentrations are increased. This behavior has been predicted successfully by the model as well.

Acknowledgments

The authors like to acknowledge 3M and DuPont for their support of this work and the National Science Foundation for its support of this work through the Presidential Faculty Fellowship to CNB (CTS-9453369) and a graduate fellowship to MDG.

Literature Cited

1.   Otsu, T.; Kuriyama, A. *Polym. J.* **1985**, *17*, 97-104.
2.   Georges, M. K.; Veregin, R. P. N.; Kazmaier, P. M.; Hamer, G. K.; Saban, M. *Macromolecules* **1994**, *27*, 7228-7229.
3.   Keoshkerian, B.; Georges, M. K.; Boils-Boissier, D. *Macromolecules* **1995**, *28*, 6381-6382.
4.   Greszta, D.; Mardare, D.; Matyjaszewski, K. *Macromolecules* **1994**, *27*, 638-644.
5.   Otsu, T.; Matsunaga, T.; Kuriyama, A.; Yoshioka, M. *Eur. Polym. J.* **1989**, *25*, 643-650.
6.   Otsu, T.; Yoshida, M.; Tazaki, A. *Makromol. Chem., Rapid Commun.* **1982**, *3*, 133-140.
7.   Lambrinos, P.; Tardi, M.; Polton, A.; Sigwalt, P. *Eur. Polym. J.* **1990**, *26*, 1125-1135.
8.   Turner, S. R.; Blevins, R. W. *Macromolecules* **1990**, *23*, 1856-1859.
9.   Dika Manga, J.; Polton, A.; Tardi, M.; Sigwalt, P. *Macromol. Reports* **1995**, *A32*, 695-703.
10.   Kannurpatti, A. R.; Lu, S.; Bunker, G. M.; Bowman, C. N. *Macromolecules* in press.
11.   Doi, T.; Matsumoto, A.; Otsu, T. *J. Polym. Sci.: Polym. Chem.* **1994**, *32*, 2911-2918.
12.   Cook, W. D. *Polymer* **1992**, *33*, 2152-2161.
13.   Anseth, K. S.; Wang, C. M.; Bowman, C. N. *Polymer* **1994**, *35*, 3243-3250.
14.   Anseth, K. S.; Wang, C. M.; Bowman, C. N. *Macromolecules* **1994**, *27*, 650-655.
15.   Anseth, K. S.; Bowman, C. N.; Peppas, N. A. *J. Polym. Sci.: Polym. Chem.* **1994**, *32*, 139-147.
16.   Marten, F.; Hamielec, A. *High conversion diffusion-controlled polymerization*; Marten, F.; Hamielec, A., Ed., 1979; Vol. 104, pp 43-70.
17.   Anseth, K. S.; Bowman, C. N. *Polym.Reaction Eng.* **1993**, *1*, 499-520.
18.   Goodner, M. D.; Lee, H. R.; Bowman, C. N. *Ind. Eng. Chem. Res.* submitted.
19.   Anseth, K. S.; Decker, C.; Bowman, C. N. *Macromolecules* **1995**, *28*, 4040-4043.
20.   Anseth, K. S.; Kline, L. M.; Walker, T. A.; Anderson, K. J.; Bowman, C. N. *Macromolecules* **1995**, *28*, 2491-2499.

Chapter 6

Light Intensity and Temperature Effect in Photoinitiated Polymerization

C. Decker, D. Decker, and F. Morel

Laboratoire de Photochimie Générale, Unité de Recherche Associée au Centre National de la Recherche Scientifique n°431, Ecole Nationale Supérieure de Chimie, Université de Haute-Alsace, 3 rue Alfred Werner, 68200 Mulhouse, France

The photoinitiated polymerization of diacrylate resins has been studied by real-time infrared (RTIR) spectroscopy on thin films exposed to UV radiation. The rate of polymerization was found to increase with the light intensity according to a nearly square root law, up to an upper limit for fluence rates superior to 100 mW cm^{-2}. This saturation effect was attributed to a fast consumption of the photoinitiator under intense illumination. The rate at which the polymer chains are growing has been evaluated from the dark polymerization profiles recorded after a short UV or laser exposure. RTIR spectroscopy has also been used to record the temperature profile of samples undergoing high-speed photopolymerization. A strong correlation was found to exist between the rate at which the temperature increases and the rate of polymerization. The temperature shows the same light intensity dependence as the reaction rate, and levels off to a maximum value under intense illumination. Photopolymerization experiments carried out at a constant temperature of 25°C clearly show that thermal runaway is not responsible for the increase of the polymerization rate observed at the beginning of the UV exposure.

UV-radiation curing has become a well-accepted technology which has found numerous industrial applications because of its distinct advantages [1-3]. One of its main characteristics is the rapidity of the process which transforms quasi-instantly the liquid resin into a solid polymer under intense illumination by a UV-source or a laser beam [4]. The polymerization rate can be finely controlled by acting on the initiation rate through the intensity of the UV radiation. It is therefore important to know how the light intensity will affect the kinetics of the polymerization, as well as the properties of photocured polymers.

Another distinct characteristic of the photocuring technology lies in the fact that the reaction is usually carried out at ambient temperature. However, in such ultrafast polymerization, the actual temperature of the irradiated sample is likely to rise to relatively high values because the heat evolved by the exothermal reaction is delivered over to the sample in a very short time. The effect of the temperature on the photoinitiated polymerization of acrylate and vinyl ether multifunctional monomers has been thoroughly investigated in the past, mainly by calorimetric analysis under isothermal conditions [5-12]. There is however little information available on the actual temperature profile of a sample undergoing high-speed photopolymerization. Such a study has been recently reported by Crofcheck et al. [13] who used a laser-induced fluorescence technique to monitor the temperature *in-situ* during the cationic photopolymerization of a divinyl ether. This elegant method presents however some intrinsic limitations : first it requires to add a luminescent temperature probe (an europium chelate) to the photosensitive formulation, with possible side effects on the polymerization kinetics, and to use a laser as light source, rather than the widely used mercury lamp ; secondly, the experiment was carried out with a 1 mm quartz capillary containing the sample, and not on thin films in contact with air, as it is done in most UV curing applications ; finally, the sample temperature could be monitored over a relatively narrow range (20 to 55°C), because of the instrumental dynamic range limitations.

The objective of the present work was to determine the influence of the light intensity on the polymerization kinetics and on the temperature profile of acrylate and vinyl ether monomers exposed to UV radiation as thin films, as well as the effect of the sample initial temperature on the polymerization rate and final degree of cure. For this purpose, a new method has been developed, based on real-time infrared (RTIR) spectroscopy [14], which permits to monitor *in-situ* the temperature of thin films undergoing high-speed photopolymerization, without introducing any additive in the UV-curable formulation [15]. This technique proved particularly well suited to addressing the issue of thermal runaway which was recently considered to occur in laser-induced polymerization of divinyl ethers [13,16].

Experimental

Materials.   The radical-type photopolymerizable formulation consisted of a mixture of hexanediol diacrylate (HDDA from UCB) and a polyurethane-diacrylate (Actilane 20 from Arkros). A bis-acylphosphine oxide (BAPO from Ciba) was used as photoinitiator at a typical concentration of 1 wt %. The cationic type photopolymerizable resin consisted of a mixture of the divinylether of triethyleneglycol (RapiCure DVE-3 from ISP) and a divinylether derivative of bis-phenol A (DVE-BPA). The cationic photoinitiator (Cyracure UVI-6990 from Union Carbide) had a composition of 50 wt % of mixed triarylsulfonium hexafluorophosphate salts and 50 wt % of propylene carbonate. The BAPO initiator was used exclusively for the acrylate polymerizations and the cationic initiator exclusively for the vinyl ether polymerizations. The chemical formulas of the various compounds used in this study are given in Figure 1.

Irradiation. The liquid resin was applied onto a 15 μm thick transparent polypropylene film by a calibrated wire-wound applicator. In most experiments a second polypropylene film was laminated onto the sample to prevent oxygen diffusion and flowing when put in vertical position. The thickness of the monomer film, typically between 5 and 25 μm, was determined accurately from its IR absorbance at 812 cm^{-1} by means of a calibration curve. The sample was inserted into a slide frame and placed in the compartment of an IR spectrophotometer, where it was exposed for a few seconds to the UV radiation of a medium pressure mercury lamp. The light intensity at the sample position was measured by a radiometer (International Light IL-390). Its value could be varied in the range 2 to 200 mW cm^{-2} by acting on the UV beam intensity through an iris diaphragm. For isothermal experiments, the laminated film was placed between two KBr crystals, and the assembly was kept at the selected temperature for 15 minutes, before being exposed to UV light.

Analysis. The polymerization reaction was followed quantitatively by monitoring in real time the decrease upon UV exposure of the distinct IR peak at 812 cm^{-1} and 820 cm^{-1} of the acrylate and vinyl ether double bonds (CH$_2$ = CH twisting), respectively. The rate of polymerization was determined from the slope of the conversion *versus* time curve, which is directly recorded by this method. The same technique (RTIR spectroscopy) was used to record the temperature profile *in-situ* of a sample undergoing high-speed photopolymerization. It is based on the fact that the intensity of the IR peak at 842 cm^{-1} of the laminated polypropylene film was found to vary linearly with the temperature over the range 20 to 100°C, in a fully reversible way. The IR absorbance value can thus be directly related to a temperature value through a calibration curve (Fig.2). Consequently, the curve recorded by RTIR spectroscopy when monitoring the variation of the absorbance at 842 cm^{-1} will accurately represent the temperature profile. In some experiments, the laminated film was applied onto a KBr crystal, because in most UV-curing applications the resin is coated onto a solid material. It should be noticed that, when the sample was clamped between two KBr crystals, the heat evolved upon the polymerization was rapidly transferred to these solids which act as a heat trap, so that the sample temperature remained essentially constant.

Influence of the Light Intensity on the Polymerization Kinetics.

The conversion *versus* time profiles were recorded by RTIR spectroscopy for a polyurethane acrylate (PUA) formulation (Actilane 20/HDDA in a 1/1 weight ratio) exposed to UV radiation of various intensities (I_0). It can be seen in Figure 3 that, after a short induction period due to the inhibitory effect of the oxygen dissolved in the formulation, the rate of polymerization rises rapidly up to a maximum value,

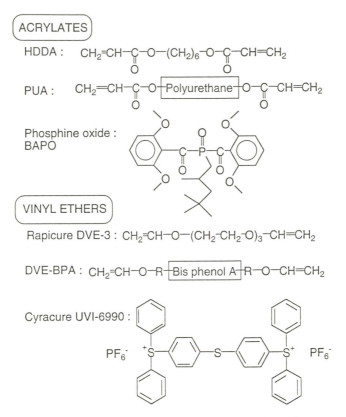

Fig. 1 Chemical formulas of the compounds used.

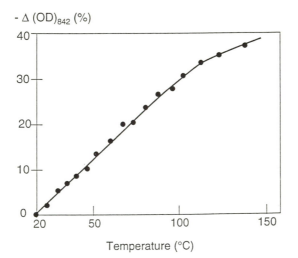

Fig.2 Variation of the polypropylene absorbance at 842 cm⁻¹ with the temperature.

$(R_p)_{max}$, which stays nearly constant up to 40% conversion. It decreases later on because of mobility restriction brought upon by gelation and solidification of the UV-irradiated material. This behavior is best illustrated in Figure 4 where the instant rate of polymerization (R_p), calculated from the slope of the curve recorded by RTIR spectroscopy, was plotted as a function of the exposure time.

The polymerization reaction was found to develop both faster and more extensively as I_0 was increased, up to a certain value above which identical RTIR curves were recorded. Consequently, the $(R_p)_{max}$ value reaches an upper limit, as shown in Figure 5 where $(R_p)_{max}$ was plotted *versus* I_0 on a logarithmic scale. The slope of the straight line obtained at low light intensities, 0.55, is close to the 0.5 value expected for a photoinitiated radical polymerization involving bimolecular termination reactions.

$$R_p = \frac{k_p}{k_t^{0.5}} (\Phi_i I_a)^{0.5} [M]$$

It is however lower than that obtained by Kloosterboer [17] in neat HDDA (0.7), which was attributed to the contribution of a first order termination process, namely radical trapping. The introduction of a flexible aliphatic polyurethane chain into the polymer network leads to an elastomeric material where mobility restrictions are expected to be less pronounced than in the hard and glassy poly(HDDA). Moreover, the cross-link density of the cured polymer is nearly half as large for the polyurethane-acrylate (PUA) as for HDDA : 4.6 mol l^{-1} compared to 8.0 mol l^{-1}. Consequently, the growing polymer radicals in PUA are less likely to become trapped during the early stage of the polymerization when $(R_p)_{max}$ is being measured (20% conversion), so that the bimolecular termination mechanism is then dominating. As the reaction proceeds and the system solidifies, radical trapping becomes an increasingly important termination process, thus making the exponent of the rate equation grow to a value close to unity [5]. It can be noticed on the R_p *versus* I_0 logarithmic plot of Figure 4 that the rate of polymerization of the vinyl ether increases linearly with the light intensity, before it levels off.

$$R_p = \frac{k_p}{k_t} \Phi_i I_a [M]$$

Such a behavior is expected in cationic polymerization where, because carbocations are not reacting among themselves, only one propagating species is involved in the termination reaction.

As the light intensity was increased above 100 mW cm^{-2}, the value of $(R_p)_{max}$ was hardly changed, as shown in Figure 5. A similar behavior was observed with the vinyl ether-based formulation, $(R_p)_{max}$ levelling off at about the same value (20 mol l^{-1} s^{-1}) as for the acrylate formulation (18 mol l^{-1} s^{-1}). Such a rate saturation effect, which has been found in nearly all of the formulations studied, is difficult to explain. It is not due to a time resolution limitation of the RTIR technique, as shown

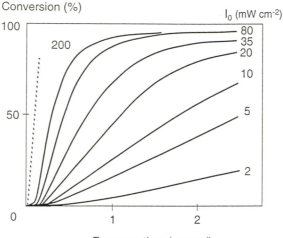

Fig.3 Influence of the light intensity on the photopolymerization of a polyurethane-acrylate (PUA) film. ---- : IR response time

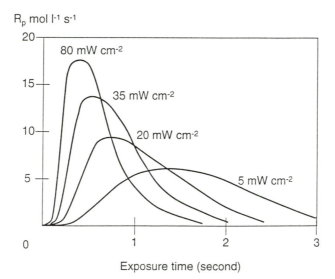

Fig.4 Variation of the rate of polymerization of a polyurethane-acrylate resin with the exposure time.

in Figure 3 (dashed curve). On the basis of the rate equation of photoinitiated radical polymerization, three factors can be considered to be responsible for the observed levelling off of R_p at high light intensities : (i) a decrease in the propagation efficiency, (ii) an increase in the termination process (in addition to the square root dependence) and, (iii) a decrease of the initiation rate ($r_i = \Phi_i\, I_a$). As the light intensity is not expected to affect the k_p and k_t rate constants, the third factor seems the most likely to be involved in the rate saturation phenomenon.

An easy way to evaluate the initiation rate, $r_i = I_0\,(1\text{-}e^{\,-2.3\,\varepsilon.l\,[PI]})\,\Phi_i$, and a possible deviation from its first-order dependence on the light intensity, is by monitoring the induction period (t_i). Under air diffusion-free conditions (laminate), a given number of radicals (N) must be generated by photolysis of the initiator to consume the O_2 molecules dissolved in the formulation, before polymerization can start :

$$N = I_0\,(1\text{-}e^{\,-2.3\,\varepsilon.l\,[PI]})\,\Phi_i\;.\;t_i = \text{constant}$$

Therefore, the reciprocal of the induction period ($1/t_i$) should vary linearly with the light intensity. Figure 6 shows that this is true, up to a certain value of the light intensity above which $1/t_i$ is levelling off, very much like $(R_p)_{max}$ does. This result suggests that the polymerization rate limitation is due, at least partly, to a decrease in the production of initiating radicals under intense illumination.

One of the possible reasons of the drop of the initiation rate could lie in the fast consumption of the photoinitiator. With a phosphine oxide photoinitiator (Lucirin TPO), as much as 75% of this compound was destroyed within 0.2 s of UV exposure at a light intensity of 400 mW cm^{-2} (Figure 7). When the light was cut off at that time (r_i=0), the polymerization was found to continue to proceed nearly as fast as upon continuous irradiation. The fact that the polymerization is only slightly faster upon continuous irradiation than in the dark suggests that r_i has already dropped to a low value when $(R_p)_{max}$ is being measured, at 20% conversion. The important post-polymerization, which lasts only a few seconds, is due to the high concentration of macroradicals that continue to polymerize in the dark.

From the cure profile recorded after a short and intense UV or pulsed laser exposure [18], one can evaluate the actual rate at which the polymer chains are growing. By calculating the ratio $R_p/[R^\bullet]$, where $[R^\bullet]$ is the number of initiating radicals generated by the UV exposure, we found that 5.10^4 acrylate double bonds have polymerized per second, for each initiating radical. From this value, the average time for the addition of one monomer unit was calculated to be 20 µs.

An increase of the light-intensity leads not only to a faster cure but also to a more complete polymerization [19]. For instance, the amount of residual unsaturation of the UV-cured acrylate polymer was shown to drop from 15 to 5% when the light intensity was increased from to 3 to 100 mW cm^{-2}. This trend can be explained by two factors : (i), an increase of the sample temperature which provides more molecular mobility and (ii), a longer time lag between conversion and shrinkage

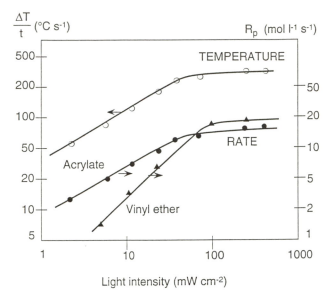

Fig.5   Influence of the light intensity on the rates of polymerization and of temperature increase of a UV-curable PUA (●,○), or phenoxy VE (s) film.

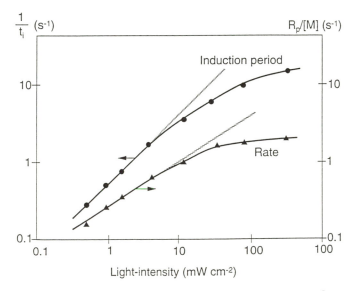

Fig.6   Light intensity dependence of the induction period (●) and the polymerization rate (▲) of a UV curable PUA film.

which generates a greater excess of free volume [17]. To quantify these effects, we have tried to measure *in-situ* the temperature of acrylate and vinyl ether monomers undergoing high-speed photopolymerization and to record in real-time the temperature profile.

Temperature profiles in photoinitiated polymerization.

During high-speed photopolymerization, the temperature of the sample increases rapidly because of the great amount of heat evolved in a very short time. Figure 8 shows the temperature profile recorded upon UV exposure of an acrylate resin by monitoring the IR peak at 842 cm^{-1} of the polypropylene film used to laminate the sample. It can be seen that the temperature starts to rise as soon as the polymerization begins. It reaches a maximum value once the reaction slows down because of gelation, and decreases later on as air cooling becomes predominant over the small amount of heat evolved in the latter stage. No significant change in the IR absorbance of the 842 cm^{-1} peak could be detected upon a 2 s UV exposure of a sample containing no photoinitiator.

As expected, an increase of the light intensity leads to both sharper temperature profiles and higher values of the maximum temperature. A linear relationship was found to exist between the rate at which the temperature rises and the rate of polymerization $(R_p)_{max}$ (Figure 9, curve a). Therefore, the speed of the temperature increase $(T_{max} - T_0)/\Delta t$ in photopolymerization of the polyurethane-acrylate is following a close to square root law with the light intensity, very much like R_p does (Fig.5). This quantity is also reaching an upper limit at high light intensities, a result which clearly shows that the rate saturation effect observed previously is not due to a time response limitation of the RTIR technique. A direct consequence of the rate and temperature saturation effects is that a polymer, photocured under different but still intense irradiation conditions, will exhibit similar physico-chemical characteristics. On the other hand, the polymer formed by exposure to low intensity UV radiation reaches a lower final conversion and exhibits a more elastomeric character. For instance, the Persoz hardness value of a polyurethane-diacrylate UV-cured at 50 mW cm^{-2} was measured to be 200 s, compared to 260 s for the sample UV-cured at 500 mW cm^{-2} with the same total energy (1 J cm^{-2}).

In Figure 9, we have also represented the variation of $(R_p)_{max}$ with the temperature at which this quantity was measured (curve b). At first sight, this nearly straight line seems to indicate that an increase of the temperature leads to a faster polymerization, a typical behavior observed in thermal runaway processes. Actually, the polymerization rate is growing here because the light intensity was increased, which leads in turn to a greater rise in temperature.

Because the temperature increase is directly related to R_p, recording the temperature profile by RTIR spectroscopy is another way to assess the performance of UV-curable formulations, and in particular the efficiency of novel

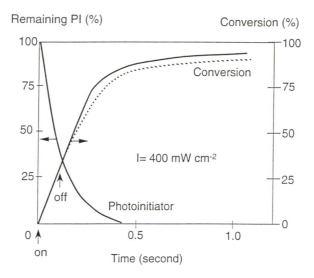

Fig.7 Polymerization and photolysis profiles of a PUA film exposed to intense UV radiation. (----- : 0.1 s exposure).

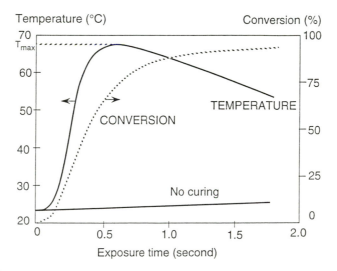

Fig.8 Temperature and conversion profiles recorded by RTIR spectroscopy upon UV exposure of a PUA film ($I_0 = 11$ mW cm^{-2})

photoinitiators. Figure 10 shows some typical temperature profiles recorded for three radical photoinitiators. They confirm the superior performance of the bis-acylphosphine oxide (BAPO) newly developed by Ciba Specialty Chemicals [20].

The maximum temperature reached upon UV exposure was found to increase with the film thickness (l), as shown in Figure 11. This result can be explained by considering that the total amount of heat evolved upon polymerization, which increases with the sample weight, is transferred to the polypropylene probe (two films of 15 μm thickness). Therefore, this method is underestimating the actual temperature reached in a free-standing film, in the absence of heat transfer, the nearest value being that obtained with thick films. The temperature data of Figure 11 were found to vary with the film thickness according to the following law : $\Delta T = \Delta T_0 \times l / (l + 30)$μm, where ΔT_0 is the temperature increase expected in the absence of the polypropylene films. For a photoinitiator concentration of 0.5%, ΔT_0 was calculated to be 150°C, a value in good agreement with that measured with a thermocouple (160°C) in a 200 μm thick film exposed to intense UV light (400 mW cm^{-2}).

Effect of the temperature on the polymerization kinetics.

One of the open questions in photoinitiated polymerization is to know whether the temperature is affecting the reaction kinetics and whether its increase during the exothermal polymerization could be responsible for a thermal runaway. To address this issue, additional photopolymerization experiments have been carried out under isothermal conditions. When the laminated sample was clamped between two KBr crystals, a flat temperature profile was obtained upon UV exposure, because of an effective heat transfer to the crystals (Fig.12). The sharp increase in the polymerization rate of the acrylate monomer was still observed when operating at a constant temperature of 25°C. Moreover, similar polymerization profiles were recorded with the acrylate sample irradiated at ambient and at 70°C (Fig.12), except for a higher final conversion at 70°C due to the enhanced molecular mobility. Therefore, thermal runaway seems unlikely to occur in the high-speed photopolymerization of these acrylic systems, under the experimental conditions used. This conclusion was confirmed by a second series of experiments performed under non-isothermal conditions where the laminated sample was placed in a slide holder, preheated at 50°C, and exposed to UV light. The polymerization profile of the acrylate formulation was very much like that recorded with the non-heated sample (Figure 13, dashed line), and the temperature profiles show a similar overall rise of the sample temperature (~ 36°C). The main results obtained in this study are summarized in Table 1. A nearly linear relationship was found to exist between the final conversion and the maximum temperature reached by the sample undergoing photopolymerization.

A similar study was carried out with vinyl ether based resins which were UV-cured isothermally at 25°C and at 70°C, in the presence of a triarylsulfonium

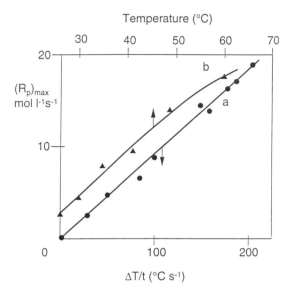

Fig.9   Variation of the rate of polymerization of PUA with the rate at which temperature rises (curve a ●) and with the temperature at which $(R_p)_{max}$ is measured (curve b ▲). (Film thickness : 10 μm).

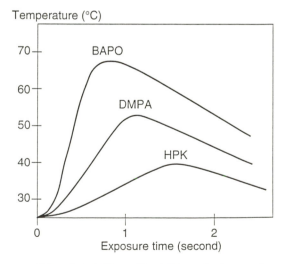

Fig. 10 Influence of the photoinitiator (1 wt %) on the temperature profile in UV curing of a PUA film. DMPA : dimethoxyphenylacetophenone ; HPK : hydroxy-phenylketone.

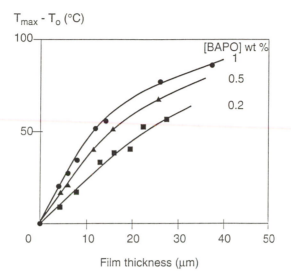

Fig.11 Variation of the temperature increase with the film thickness upon UV-curing of a PUA film (I_0 = 11 mW cm^{-2}).Photoinitiator concentration : 0.2% (■) ; 0.5% (▲) ; 1% (●).

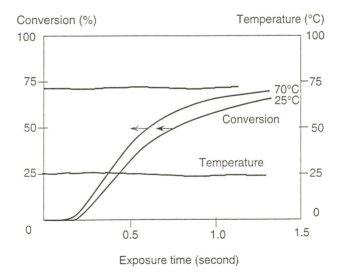

Fig.12 Isothermal photopolymerization of PUA film irradiated at 25°C or 70°C.

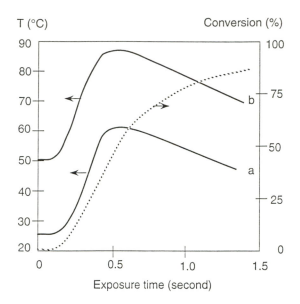

Fig.13 Temperature profile of a UV-curable PUA irradiated at an initial
temperature of 25°C (curve a) or 50°C (curve b). ($I_0 = 11$ mW cm^{-2})
---- : conversion profile at 50°C.

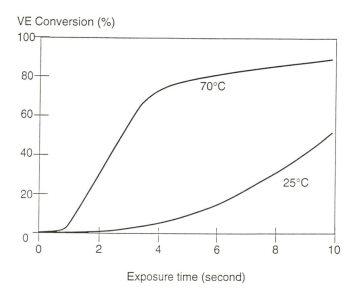

Fig.14 Influence of the temperature on the isothermal photopolymerization of a
phenoxy-divinyl ether. [Cyracure UVI-6990] = 4 wt %. $I_0 = 17$ mW cm^{-2}

Table 1 : Influence of the temperature on the photopolymerization of a polyurethane-acrylate resin (Light intensity : 11 mW cm^{-2})

Irradiation conditions	Isothermal		Non-isothermal	
Initial temperature (°C)	25	70	25	50
Polymerization rate (mol l^{-1}s^{-1})	8	10	9	10
Final conversion (%)	75	90	87	94
Maximum temperature (°C)	25	70	60	87

photoinitiator. In contrast to acrylate resins, here a marked increased of the polymerization rate was observed when the bisphenol A-divinyl ether resin was UV irradiated at 70°C instead of at 25°C, as shown in Figure 14. Such a strong temperature effect, which seems to point toward a thermal runaway process, is actually due to a viscosity effect. Indeed, the rate of polymerization of the vinyl ether was found to increase sharply as the viscosity was decreased by lowering the ratio BPA-DVE/DVE-3. Figure 15 shows the variation of the reactivity ($R_p/[M]_0$) of the vinyl ether formulation as a function of the viscosity. Because R_p varies with the vinyl ether concentration, the ratio $R_p/[M]_0$ must be used for a reliable evaluation of the viscosity effect. It can be seen that the data obtained at 70°C (triangles) fall perfectly well on the curve $R_p/[M]_0 = f(\eta)$ relative to the UV curing at 25°C.

Further evidence showing that the temperature *per se*   is not influencing significantly the photoinitiated cationic polymerization of vinyl ether monomers was obtained by studying the UV-curing of DVE-3. The viscosity of this highly fluid monomer is essentially the same at 25°C and 70°C. Very similar polymerization profiles were recorded by RTIR spectroscopy when DVE-3 was photopolymerized isothermally at either 25°C or 70°C, as shown in Figure 16. The kinetic data of this study are summarized in Table 2. They confirm that, like for acrylate-based systems, thermay runaway is not occurring during the photoinitiated cationic polymerization of the vinyl ether monomers studied.

Table 2 : Influence of the temperature on the cationic photopolymerization of the divinyl ether of triethyleneglycol (DVE-3) (I = 5 mW cm^{-2}). Film laminated between 2 KBr crystals

Temperature (°C)	25	40	60	70
Induction period (s)	1.7	1.6	1.7	1.7
Reactivity : $R_p/[M]_0$(s^{-1})	1.2	1.3	1.2	1.3
Final conversion (%)	92	94	93	93

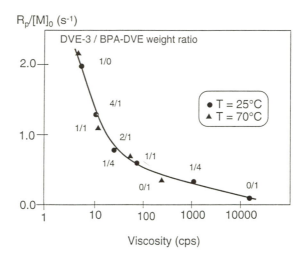

Fig.15 Viscosity dependence of the polymerization reactivity of a UV-curable phenoxy-divinyl-ether ($I_0 = 17$ mW cm^{-2}).

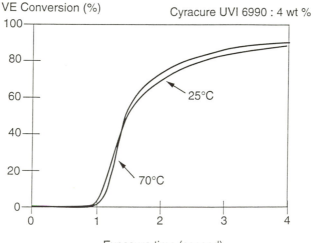

Fig.16 Influence of the temperature on the photopolymerization of divinylether of triethylene glycol. ($I_0 = 17$ mW cm^{-2})

Conclusion

The temperature reached by a monomer undergoing photopolymerization plays a key role on the reaction kinetics, in particular on the ultimate degree of conversion and therefore on the physico-chemical properties of the UV-cured polymer. It is strongly dependent on the formulation reactivity, the film thickness, as well as on the light intensity.

A novel method, based on real-time infrared spectroscopy, has been developed to monitor the temperature in thin films undergoing high-speed photopolymerization. The temperature probe is a polypropylene film, laminated onto the sample, which exhibits a temperature sensitive IR peak at 842 cm^{-1}. The sharp rise of the sample temperature upon UV exposure of diacrylate monomers was shown to be directly related to the polymerization process and to depend on chemical and physical parameters, such as the photoinitiator concentration and the light intensity. For the acrylate and vinyl ether resins studied, thermal runaway is not responsible for the large increase in the polymerization rate occurring soon after the beginning of the UV exposure. This new technique could find applications in any system undergoing ultrafast polymerization in which non-intrusive measurements of the temperature is required, and in particular in clear and pigmented thin films, as well as in composite materials.

References
1. *Radiation Curing of Polymeric Materials,* Hoyle C.E, Kinstle J.F. (Eds), ACS Symp.Series 417, ACS Washington DC, **1990**.
2. Radiation Curing-Science and Technology, Pappas S.P. (Ed.), Plenum Press, New York, **1992**.
3. Decker C., *Prog.Polym.Science*, **1996**, *21*
4. Decker C., *Acta Polymer*, **1994**, *45*, 333
5. Tryson G.R., Shultz A.R., *J.Polym.Sci., Polym.Phys.Ed.*, **1979**, *17*, 2059 .
6. Crivello J.V., Lee J.L., Coulon D.A., *J.Rad.Curing*, **1983**, *10*, 6 .
7. Bessers H.J.L., Kloosterboer J.G., *Polym.Bull.*, **1980**, *2*, 201
8. Hoyle C.E., Hensel R.D., Grubb M.B., *J.Polym.Sci., Polym.Chem.Ed.*, **1984**, *22*, 1865 .
9. Bair H.E., Blyer L.L., *Proc. 14th NATAS* p.392, **1985**.
10. Appelt B.K., Abadie M.J.M., *Polym.Eng.Sci.,* **1985**, *25(15)*, 931.
11. Broer D.J., Boven J., Mol. G.N., Challa G., *Makromol.Chem.*, **1989**, *190*, 2255.
12. Nelson E.W., Jacobs J.L., Scranton A.B., Anseth K.S., Bowman C.N., *Polymer*, **1995**, *36*, 4651.
13. Crofcheck C.L., Nelson E.W., Jacobs J.L., Scranton A.B., *J.Polym.Sci., Polym.Chem.Ed.*, **1995**, *33*, 1735.
14. Decker C., Moussa K., *Makromol.Chem.*, **1988**, *189*, 2381.
15. Decker C., Decker D., Morel F., *Polym.Mater.Sci.Eng.*, **1996**, *74*, 350.
16. Nelson E.W., Carter T.P., Scranton A.B., *Macromolecules*, **1994**, *27*, 1013.

17.   Kloosterboer J.G., *Adv.Polym.Sci.*, **1988**, *84*, 1

18.   Decker C., *Europ.Polym.Paint Col.J.*, **1992**, *182*, 383.

19.   Decker C, Elzaouk B., Decker D., *J.Macromol.Sci. Pure Appl.Chem.*, **1996**, *A33*, 173 .

20.   Misev L., *Aspects of Photoinitiation*, Paint Res.Assoc.Teddington, England, **1993**, p.299.

Cationic, Charge-Transfer, and Thiol-ene Systems

Chapter 7

Structure and Reactivity Relationships in the Photoinitiated Cationic Polymerization of 3,4-Epoxycyclohexylmethyl-3′,4′-epoxycyclohexane Carboxylate

James V. Crivello and Ulrike Varlemann[1]

Department of Chemistry, Rensselaer Polytechnic Institute, Troy, NY 12180–3590

An investigation of the reactivity of the cycloaliphatic diepoxide, 3,4-epoxycyclohexyl 3',4'-epoxycyclohexane carboxylate (I) in photoinitiated cationic polymerization was carried out with the aid of model compounds to elucidate the effects of structure on the reactivity of this biscycloaliphatic epoxide. It was shown that the rate of photoinitiated cationic polymerization of this monomer is greatly affected by the presence of the ester group. Molecular modeling studies indicate that the ester carbonyl group can interact either intra- or intermolecularly with the growing oxonium cation to give dioxacarbenium ions. These latter species are both more sterically hindered and less reactive than their oxiranium cation precursors and undergo propagation at a considerably reduced rate. It was further demonstrated that ether groups present either in the monomer or generated on polymerization are also rate retarding. Based on these studies, a series of biscycloaliphatic epoxide monomers with higher reactivity than I were prepared.

Photoinitiated cationic polymerizations of epoxides have been the subject of considerable research in this laboratory during the past several years (1,2). As the commercial applications of the photoinitiated cationic polymerizations of epoxides have multiplied there has arisen a greater need to understand the influence of the structure on the reactivity of these monomers in order to maximize their rates of photopolymerization. During our early investigations in this area we noted that there appeared to be a wide range of reactivity between structurally diverse epoxides (3). It was observed that among the commercially available epoxy monomers, the biscycloaliphatic epoxy monomer, 3,4-epoxycyclohexylmethyl 3',4'-epoxycyclohexane carboxylate (I) is among the most reactive and, in addition, has a low order of toxicity. Consequently, this monomer has been extensively employed in many commercial applications involving photoinitiated cationic polymerization (UV curing).

[1]Current address: Coatings and Inks Division, Henkel Corporation, 300 Brookside Avenue, Ambler, PA 19002

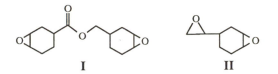

I   **II**

At the same time, when **I** is compared with 4-vinylcyclohexene diepoxide (**II**), it was observed that **II** is considerably more reactive. The reasons for the difference in the reactivity between **I** and **II** were not immediately apparent and, therefore, a study of the reactivity of **I** was undertaken with the aid of model compounds.

EXPERIMENTAL

Materials

1,2,3,4-Tetrahydrobenzyl alcohol (($\pm$)3-cyclohexenene-1-methanol) and 30% aqueous hydrogen peroxide were purchased from Fluka, AG. 3-Cyclohexene-1-carboxylic acid and cis-4-cyclohexene-1,2-dicarboxylic acid were used as purchased from Lancaster Chemical Co. Methyl iodide, acetic anhydride, Oxone® (potassium peroxymonosulfate), Aliquot 336® (methyl tri-n-octylammonium chloride), sodium tungstate dihydrate and N,N-dimethylaminopyridine (DMAP) were purchased from Aldrich Chemical Co. and used as received. 3,4-Epoxycyclohexylmethyl 3',4'-epoxycyclohexane carboxylate (ERL 4221) and 4-vinylcyclohexene dioxide were used as purchased from the Union Carbide Corp. (4-n-Octyloxyphenyl)phenyliodonium hexafluoroantimonate used as a photoinitiator was prepared by a procedure described previously (4).

Routine infrared spectra were obtained on a Buck Scientific model 500 spectrometer. ^{1}H-NMR spectra were obtained using a Varian XL-200-MHz Spectrometer at room temperature in CDCl3 solvent. The chemical shifts are reported relative to tetramethylsilane used as an internal standard. Mass spectra were recorded on a Shimadzu Corporation Model GCMS-QP1000 using chemical ionization techniques and employing isobutane as the reagent gas. Elemental Analyses were performed by Quantitative Technologies Inc., Whitehouse, N.J.

Synthesis of Model Compounds

The following are typical of the synthetic procedures employed for the preparation of model compounds **IV-VIII**.

Methyl 3-Cyclohexene-1-carboxylate

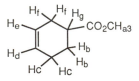

Into a 100 mL round bottom flask fitted with a magnetic stirrer, and a condenser were placed 10.09 g (0.08 mol) of 3-cyclohexene-1-carboxylic acid, 12.81 g (0.40 mol) of methanol and 1.56 g (0.016 mol) of concentrated sulfuric acid. The reaction mixture was heated under reflux for 5 hr in an oil bath. After this time, the reaction mixture was diluted with water and then extracted several times with ether. The organic phase was neutralized by washing with a dilute solution of sodium carbonate, then with distilled water and finally dried over anhydrous magnesium sulfate. After filtration, the solvent was removed using a rotary evaporator and the crude product distilled under high vacuum. There were obtained 8.93 g (80% theory) of the desired product, methyl 3-cyclohexene-1-carboxylate, as a colorless oil with a b.p. of 50-51°C at 0.5 mm Hg.

IR (neat), NaCl (cm^{-1}) 3030 b, 2950 s, 2850 s, 1780 s, 1665 m, 1646 s, 1436 m, 652s.

^{1}H-NMR (CDCl$_3$) d (ppm) 1.5-1.8 (m, 1H, H$_g$); 1.92-2.2 (m, 6H, H$_{b,c,f}$); 3.65 (s, 3H, H$_a$); 5.65 (m, 2H, H$_{d,e}$).

MS: parent peak m/z 140.

Elemental Analysis Calc. for C$_8$H$_{12}$O$_2$: %C, 68.55; %H, 8.63; %O, 22.82. Found: %C, 67.66; %H, 8.50; %O, 24.12.

Methyl 3,4-Epoxycyclohexane-1-Carboxylate (**IV**)

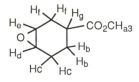

Added to a 100 mL round bottom flask were 13.36 g (24 mmol) of methyl 3-cyclohexene-1-carboxylate, 15 mL of CHCl$_3$, 0.19 g (0.48 mmol) of Aliquat$^®$ 336, 12.25 g of 8% aqueous hydrogen peroxide, 0.39 g (1.2 mmol) of sodium tungstate dihydrate and 0.75 g of 40 w/v% phosphoric acid. The flask was equipped with a magnetic stirrer and a reflux condenser and the reaction mixture was heated at 65°C for 16 h. Then the aqueous and organic phases were separated by means of a separatory funnel and the organic phase washed several times with distilled water. The organic phase was then dried over anhydrous sodium sulfate. After filtration, the solvent was removed on a rotary evaporator and the resulting oil distilled under high vacuum. There were obtained 1.87 g (50% theory) of a colorless liquid epoxide, **IV**, having a b.p. of 37°C at 0.1 mm Hg.

^{1}H-NMR (CDCl$_3$) d (ppm) 1.2-2.6 (m, 7H, H$_{b,c,f,g}$); 3.15 (s, 2H, H$_{d,e}$); 3.65 (s, 3H, H$_a$).

Elemental Analysis Calc. for C$_8$H$_{12}$O$_3$: %C, 61.52; %H, 7.74; %O, 30.73. Found: %C, 61.13; %H, 7.87; %O, 30.06.

3,4-Epoxycyclohexanemethanol (**VIII**)

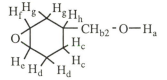

A stirred solution of 5.00 g (44.58 mmol) of 3-cyclohexene-1-methanol, 556 mL acetone and 1.39 mL of an aqueous phosphate buffer (pH = 7.2) was cooled to 0°C and then 34.25 g (55.72 mmol) of Oxone$^®$ (0.4 M solution in water) was added. The phosphate buffer was prepared by dissolving 1.77 g (13.04 mmol) KH$_2$PO$_4$ and 6.48 g (45.64 mmol) Na$_2$HPO$_4$ in distilled water and then diluting to give a total volume of 1.5 L. The pH during the reaction was kept at 7.2 by adding 3N NaOH dropwise. The mixture was allowed to warm to room temperature and stirring was continued for an additional 12 h. The reaction mixture was transferred to a separatory funnel and extracted several times with methylene chloride. The methylene chloride extracts were combined, dried over anhydrous magnesium sulfate and then filtered. The solvent was removed under reduced pressure leaving 3,4-epoxycyclohexene-1-methanol (**VIII**) as a colorless oil.

^{1}H-NMR (CDCl$_3$) d (ppm) 0.8-2.25 (m, 7H, H$_{c,d,g,h}$); 3.05-3.25 (s,s, 3H, H$_{a,e,f}$); 3.4-3.5 (d, 2H, H$_b$).

Elemental analysis calculated for C$_7$H$_{12}$O$_2$: %C, 65.59; %H, 9.44. Found: %C, 65.20; %H, 9.50.

Synthesis of Novel Biscycloaliphatic Epoxides

The synthetic procedures given below are typical for those used for the synthesis of difunctional epoxycyclohexyl monomers **XXIIIa-e**.

1,4-Bis(3-cyclohexenylmethoxy)butane (**XXIIa**) Into a three necked round bottom flask equipped with a reflux condenser, nitrogen inlet, Normag bubbler and a magnetic stirrer were added 33.64 g (0.30 mol) of 3-cyclohexene-1-methanol, 21.59 g (0.10 mol) of 1,4-dibromobutane, 20 g (0,50 mol) of powdered NaOH and 80 mL of toluene. The mixture was stirred at room temperature for 2 h and then heated to 80°C and maintained at this temperature for 90 h. After cooling, the reaction mixture was transferred to a separatory funnel and the organic and aqueous layers separated. The aqueous layer was extracted with toluene and the organic layers combined and washed several times with water. The toluene phase was dried over anhydrous sodium sulfate, filtered and the solvent removed using a rotary evaporator. The product, **XXIIa**, was obtained as an oil and was purified by fractional distillation under high vacuum. A yield of 29% (8.19 g) of the pure product having a b.p. of 123-124°C at 0.1 mm Hg was obtained.

In the synthesis of the higher analogues (**XXIIc-e**), purification was achieved by high vacuum stripping of the starting materials and monoadduct from the desired diolefin product.

1,4-Bis(3,4-epoxycyclohexylmethyloxy)butane (**XXIIIa**) A 50 mL single necked round bottom flask equipped with a magnetic stirrer, addition funnel and a reflux condenser was first cooled in an ice bath for 10 minutes. It was then charged with 2.50 g (8.98 mmole) of **XXIIa** and 50 mL of dichloromethane and this mixture cooled to 0°C. To this mixture there was added in small portions over 10 minutes with stirring 3.11 g (18.03 mmole) of m-chloroperoxybenzoic acid (m-CPBA). Throughout the addition, the reaction mixture was maintained at a pH of 8 by the dropwise addition of a solution of 0.1 M NaHCO$_3$. After stirring for an additional period of 5 h, the mixture was again cooled to 0°C and then a second 3.11 g portion of m-CPBA was added in the same manner. After further stirring for a period of 5 h, the reaction mixture was transferred to a separatory funnel and the phases separated. The dichloromethane layer was retained and washed several times with water and finally dried over anhydrous sodium sulfate. The solvent was removed on a rotary evaporator and the final product placed under high vacuum overnight to remove traces of solvent and water. There were obtained 1.97 g (70% yield) of diepoxide **XXIIIa** as a colorless oil.

Real Time Infrared (RTIR) Measurements

RTIR measurements were carried out as described a previous communication (5).

Molecular Modeling

Structures of various dioxacarbenium and oxiranium intermediates were modeled using CAChe MOPAC molecular orbital software along with a Macintosh IIci workstation. The conformations and energies of the structures were first computationally minimized and then the heats of formation were calculated.

RESULTS AND DISCUSSION

Synthesis of Model Compounds

Before undertaking a study of the reactivity of **I**, a few general observations concerning its structure should be noted. First, commercially available **I** is not a pure compound, rather, due the presence of four independent chiral centers in the molecule, it consists of a mixture of sixteen stereoisomers. Most apparent is the observation that some of these isomers possess epoxide groups which are either *cis* or *trans* to the ester groups lying across the cyclohexane ring. It is also worth pointing out, that the molecule is clearly unsymmetrical with different functional groups attached to each epoxycyclohexyl group and consequently, the reactivities of the two epoxy groups would be expected to be different.

A series of model compound epoxides (**III-VIII**) representing various portions of the **I** molecule were prepared and are listed below. The methods for their preparation are outlined in equations 1-4 of Scheme 1.

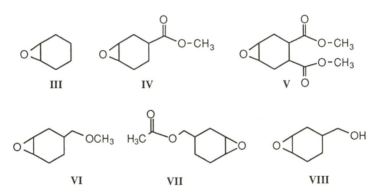

The simplest model compound is cyclohexene oxide **III**. Monomers **IV**, **V** and **VII** represent different aspects of the ester portion of **I**, while monomers **VII** and **VIII** reflect aspects of both the monomer **I** and the polymer which is formed by cationic ring-opening polymerization. Monomers **IV-VII** were prepared using a phase transfer catalyzed epoxidation based on the method of Venturello and D'Aloisio (6) and employed previously in this laboratory (7). This method was not effective for the preparation of monomer **VIII**. In this specific case (equation 4), epoxidation using Oxone® (potassium monoperoxysulfate) was employed.

Cationic Photopolymerization

To evaluate the reactivity of model compounds **III-VIII** in photoinitiated cationic polymerization, we have employed real-time infrared spectroscopy (RTIR). Thin film samples of the model compounds containing 0.5 mol% of (4-n-octyloxyphenyl)phenyliodonium SbF_6^- as a photoinitiator were irradiated in a FTIR spectrometer at a UV intensity of 20 mW/cm². During irradiation, the decrease in the absorbance of the epoxy ether band at 860 cm^{-1} was monitored.

Figure 1 shows a RTIR comparison of the photopolymerizations of the above seven model compounds. The initial slopes of the curves were calculated and are shown in Table 1. Based on this data, the monomers may be ranked in the following order with respect to their rates of polymerization.

Scheme 1

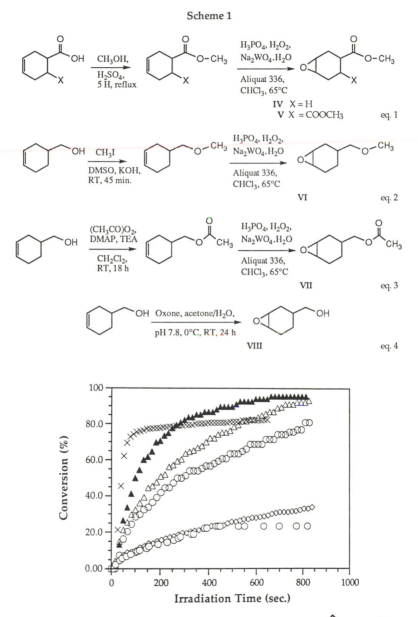

Figure 1. RTIR study of the the photopolymerizations of (◇), **I**; (△) **IV**; (⊙), **V**; (▲), **VI**; (○), **VII**; (X), **VIII** in the presence of 0.5 mole% (4-n-octyloxyphenyl)phenyliodonium SbF$_6^-$.

Table 1

RTIR Studies of Model Compounds[*]

Structure	Notation	$R_p/[M_o]$[†]
	I	0.17
	IV	0.23
	III	2.13
	VI	0.48
	VII	0.24
	VIII	1.66
	V	0.13

[*]Studies conducted in the presence of 0.5 mol% (4-n-octyloxyphenyl)phenyl-iodonium SbF_6^-. [†]Calculated according to the equation:
$R_p/[M_o] = [(conversion_{t1}) - (conversion_{t2})]/(t_1 - t_2)$.

To this point, only intramolecular reactions in which either the ester or ether group are position *trans* to the oxiranium ion have been considered. The corresponding *cis* isomers cannot undergo backside attack on the oxiranium ion, however, the ester and ether groups can participate in stabilization of the protonated epoxide species by hydrogen bonding with the epoxide oxygen to form intermediates such as **XV** and **XVI**. Calculations show (Table 2) that both of these intermediates are more stable than a protonated or alkylated oxiranium ion.

Similarly, intermolecular interactions of ester and ether groups with growing oxiranium ions to form intermediates analogous to those shown in Table 2 would be expected to take place. In each case, the effects would be predicted to diminish the reactivity of the growing cation.

Mechanistic Interpretation

On the basis of the above model compound studies, the reactivity of **I** can now be more clearly understood. The nucleophilic ester groups in **I** can interact intramolecularly with oxiranium ions formed by attack at either of the two alkylated epoxide groups to form dioxacarbenium ions **XVIII**, **XIX** and **XXI** which are less reactive than oxiranium ions **XVII** or **XX**. This is shown in Scheme 2.

Scheme 2

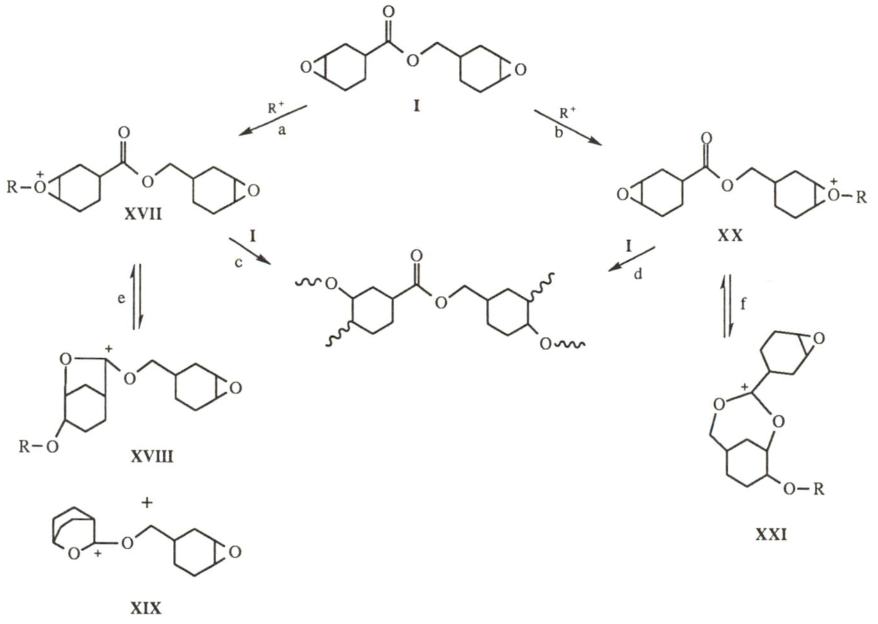

Polymerization is slowed since the dioxacarbenium species lie in deeper energy wells than the corresponding oxiranium ions and consequently, react more slowly. A similar mechanistic scheme can be written for the intermolecular reaction between the ester group of one molecule of **I** and the oxiranium ion of a second molecule of **I**.

Table 2
Calculated Heats of Formation for Proposed Intermediates

Structure	ΔH_f (kcal/mol)	Structure	ΔH_f (kcal/mol)
IX	37.8	XIII	80.5
X	29.2	XIV	76.9
XI	26.6	XV	54.5
XII	104.3	XVI	102.1

III > VIII > VI > VII > IV > I > V

Monomer **VIII** bearing both a hydroxyl group and an epoxycyclohexyl group shows surprisingly high reactivity.   It should be pointed out that because of the hydroxyl group, this monomer undergoes polymerization by an activated monomer mechanism rather than by a conventional cationic epoxide ring-opening mechanism (8,9). The kinetics of the two polymerization reactions are different.

Considering the above data, one can conclude that the presence of ester carbonyl groups in a monomer diminishes the rate of polymerization of the epoxy groups in the same molecule as compared to unsubstituted cyclohexene oxide (**III**). Similarly, comparing monomers **I** and **IV** with monomer **V**, it can be seen that the introduction of a second ester carbonyl group in this latter molecule further slows the polymerization rate. The presence of the ether groups in the monomer **VI** also appears to adversely affect the polymerization rate of epoxycyclohexanes although not as much as an ester group.

The above observations with respect to the reactivities of monomers **IV-VI** can be explained by postulating a direct interaction of the carbonyl and the ether functional groups with the propagating cationic center. This can occur by either an inter- or intramolecular process. As shown in equation 5, intramolecular backside attack by the ester carbonyl group of the d,l-*trans* **IV** isomers at either carbon of the protonated or alkylated epoxy group gives rise to bicyclic dioxacarbenium ions **IX** and **X**.

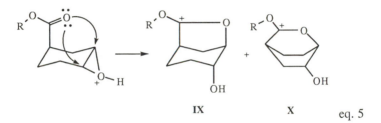

IX   X   eq. 5

There is considerable literature precedent for this reaction. In particular, Fotsch and Chamberlin (10) have reported that open chain γ,δ, δ,ε and ε,ζ-epoxy ketones and esters undergo cyclization in the presence of acids to form the corresponding dioxacarbenium ions. In addition, molecular orbital calculations were conducted to determine the heats of formation of the intermediates **IX** and **X**. Data from these calculations are given in Table 2. These calculations suggest that 1,6-attack (**X**) is preferred over 1,5-attack (**IX**). Furthermore, the ΔH_f values for both of these two intermediates are considerably lower than that calculated for a strained oxiranium ion intermediate such as **XII**. Accordingly, it would be predicted that the dioxacarbenium ions **IX** and **X** possess similarly lower activity than oxiranium ion intermediates (such as **XII**) which are the propagating species in conventional cationic epoxide polymerizations. It is interesting to note that the carbonyl group of ester model compound **VII** can bridge to form a seven membered ring intermediate **XI** which also has a low ΔH_f value (26.6 kcal/mol). Similar bicyclic intermediates can be written for diester-containing model compound **V**. In the case of ether-containing monomer **VI**, bridged 5 and 6 membered oxonium ions **XIII** and **XIV** can be formed by nucleophilic attack of the ether group on the propagating oxiranium ion. Since these intermediates also have relatively lower calculated ΔH_f values, these species would be preferentially formed and also have a deaccelerating effect on the overall rate of polymerization.

Synthesis of Novel Monomers

The above mechanistic study suggested that to construct monomers more reactive than **I**, the monomer should bear highly reactive epoxycyclohexyl groups, but should not contain ester or other nucleophilic groups. Accordingly, we initiated a program to prepare such a novel series of difunctional epoxide monomers. Shown in equations 6 and 7 is the synthetic method which was employed.

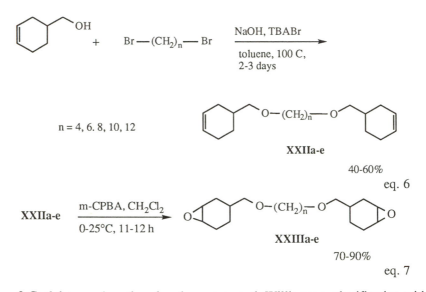

n = 4, 6. 8, 10, 12

XXIIa-e

40-60%

eq. 6

XXIIa-e $\xrightarrow[\text{0-25°C, 11-12 h}]{\text{m-CPBA, CH}_2\text{Cl}_2}$

XXIIIa-e

70-90%

eq. 7

3-Cyclohexene-1-methanol undergoes smooth Williamson etherification with α,ω-dibromoalkanes in the presence of base and a phase transfer catalyst. The resulting biscyclohexenyl ethers, **XXIIa-e**, were subsequently treated with m-chloroperoxybenzoic acid to give the desired diepoxide monomers, **XXIIIa-e**. Table 3 gives the characteristics of these monomers.

It was found that the reactivity of monomers **XXIIIa-e** depends on the length of the alkylene chain. When the length of the chain is short (4-8 carbon atoms), the reactivity is high and exceeds that of diepoxide monomer **I**. For example the RTIR curves for **I** and **XXIIIc** are shown in Figure 2. However, when the alkylene chain is lengthened (10-12 carbons), the polymerization rates decrease. This may be attributed to several factors specific to these difunctional monomers including: the decreased solubility of the photoinitiators, a lower statistical probability of intramolecular reaction at the two epoxide ends of the molecule and also to the very high viscosity of monomers **XXIIId** and **XXIIIe**. 4-n-Octyloxyphenylphenyliodonium SbF$_6^-$ is only soluble in the warm monomers and may begin to precipitate during polymerization. In addition, the high viscosity of the monomers has a retarding effect on the rate by slowing the diffusion of unreacted monomer into the propagating sites located within the swollen three dimensional network of the already formed polymer.

Table 3

Characteristics of 1,4-Bis(3,4-Epoxycyclohexylmethyloxy)alkanes

$$O \overset{}{\underset{}{\bigcirc}} \hspace{-0.5em} \underset{}{\diagdown} CH_2-O-CH_2-(CH_2)_{n-2}-CH_2-O-CH_2 \overset{}{\underset{}{\diagup}} \hspace{-0.5em} \bigcirc O$$

Notation	n	Yield (%)	Elemental Analysis	%C	%H
XXIIIa	4	70	Calc:	69.65	9.74
			Fnd:	69.29	9.67
XXIIIb	6	60	Calc:	70.98	10.12
			Fnd:	71.03	9.96
XXIIIc	8	77	Calc:	72.09	10.45
			Fnd:	71.84	10.26
XXIIId	10	91	Calc:	73.05	10.73
			Fnd:	72.97	10.82
XXXIIIe	12	79	Calc:	73.89	10.97
			Fnd:	73.25	10.68

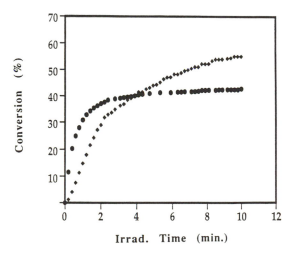

Figure 2. RTIR comparison between the photopolymerizations of (◆), **I** and (●), **X X I I I c** in the presence of 0.5 mole% (4-n-octyloxyphenyl)phenyliodonium SbF$_6$-.

It should also be noted that there is a considerable difference in the extent of the final conversion between the monofunctional monomers shown in Figure 1 and those obtained for the difunctional monomers given in Figure 2. In the case of the monofunctional monomers, linear polymers are formed and the conversions reached are typically high. In contrast, conversions of 50-60% are obtained for the difunctional monomers **I** and **XXIIIc**. In these latter cases, the final conversion is limited due to the formation of rigid, crosslinked networks which immobilizes a substantial portion of the functional groups and prevents them from further reaction. Higher conversions can be reached in these systems by raising the temperature above the T_g of the crosslinked matrix.

CONCLUSIONS

The reactivity of **I** in photoinitiated cationic polymerization is due to several factors associated with the structure of this monomer. Most importantly, the presence of the ester groups in **I** which can interact with oxiranium ions generated at either of the two epoxide groups both intra- and intermolecularly produces dioxacarbenium ions of reduced activity in the propagation reaction. Taking this into account, a series of diepoxides were prepared which did not possess ester groups. Some of these monomers show enhanced reactivity as measured by RTIR in photoinitiated cationic polymerization compared to **I**.

LITERATURE CITED

1. J.V. Crivello and J.H.W. Lam, ACS Symp. Ser., 114, *Epoxy Resin Chemistry,* R.S. Bauer, editor, Am. Chem. Soc., Washington, D.C., 1979, p. 1.
2. J.V. Crivello and J.L. Lee, ACS Symp. Ser., 417, *Radiation Curing of Polymeric Materials,* C.E. Hoyle and J.F. Kinstle, editors, Am. Chem. Soc., Washington, D.C., 1990, p. 398.
3. J.V. Crivello, J.H.W. Lam and C.N. Volante, *J. Rad. Curing,* **1977**, 2, 2.
4. J.V. Crivello, and J.L. Lee *J. Polym. Sci., Polym. Chem. Ed.,* **1989**, 27, 3951.
5. J.V. Crivello and K.D. Jo, *J. Polym. Sci., Polym. Chem. Ed.,* **1993**, 31, 1483.
6. C. Venturello and R. D'Aloisio, *J. Org. Chem.,* **1988**, 53, 1553.
7. J.V. Crivello and R. Narayan, *Chemistry of Materials,* **1992**, 4(3), 692.
8. E.J. Goethals, G.G. Trossaert, P.J. Hartmann and R.R. De Clerq, *Polymer Preprints,* **1993**, 34(1), 205.
9. S. Penczek and P. Kubisa, In *Ring-Opening Polymerization,* D.J. Brunelle, editor, Hanser Verlag, Munich, 1993, p. 19.
10. C.H. Fotsch and A.R. Chamberlin, *J. Org. Chem.,* **1991**, 56, 4141.

Chapter 8

Effect of Viscosity on the Rate of Photosensitization of Diaryliodonium Salts by Anthracene

S. K. Moorjani, B. Rangarajan, and Alec B. Scranton[1]

Department of Chemical Engineering, Michigan State University, A202
Engineering Building, East Lansing, MI 48824–1226

We have examined the effects of viscosity on the rate of photosensitization of diaryliodonium salts by anthracene. Fluorescence spectroscopy was used to monitor the photosensitization reaction in propanol/glycerol solutions of differing viscosities. As the viscosity was increased, the rate of photosensitization decreased in a manner qualitatively described by the Smoluchowski-Stokes-Einstein equation for bimolecular elementary reactions, coupled with numerical solution of the photophysical equations. Based on these studies, we would expect the viscosity change during polymerization to have a profound effect on the rate of photosensitization due to a marked decrease in the rates of the diffusion-controlled bimolecular reactions. Therefore, the rate of photosensitization may become extremely low after a certain degree of polymerization with further reaction arising primarily from propagation of existing active centers.

The use of light to induce polymerization reactions leads to many advantages. For example, photopolymerizations allow rapid reaction rates and provide superior spatial and temporal control over the initiation reaction. In addition, photopolymerizations require considerably less energy than conventional thermal polymerizations in which the entire reaction system is raised to elevated tempertures.[1] Finally, since photopolymerizable formulations are typically solvents-free, these reactions may lead to a decrease in the emission of volatile organic chemicals. This set of advantages has lead to tremendous growth in the use of photopolymerizations for a variety of applications including coatings, dental materials, and a host of emerging high-tech and electronic applications such as coatings on optical fibers, replication of optical disks, and fabrication of integrated circuits.

Cationic photopolymerizations of epoxides and vinyl ethers offer tremendous potential in the area of high-speed, solvent-free curing of films and coatings. The polymers formed exhibit excellent clarity, adhesion, abrasion, and chemical

[1]Corresponding author

resistance.[2-4]   In addition, UV-initiated cationic polymerizations offer several processing advantages. These reactions are not inhibited by oxygen and do not require an inert atmosphere. In contrast to free radicals, the cationic active species do not terminate by combination; therefore, cationic polymerizations may proceed long after the irradiation has ceased.[5,6]   Finally, several monomers are available which exhibit low vapor pressures, low viscosities and low toxicity, but polymerize rapidly to form films exhibiting excellent properties.[7-14]   These advantages have resulted in a surge in research devoted to characterizing the mechanism and kinetics of these reactions, as well as the physical properties of the polymeric films obtained.

Interest in UV-initiated cationic photopolymerizations was fueled by the development of two classes of thermally stable photoinitiators: diaryliodonium and triarylsulfonium salts.[3,15]   While these salts are most efficiently initiated by deep-UV (225-275 nm) light, the effective initiating wavelength may be expanded by the use of photosensitizers or co-initiators such as anthracene, isopropyl thioxanthone, or chloro-propoxy thioxanthone which shift the effective spectral window to well above 300 nm.   This near-UV region of the spectrum is particularly convenient for initiating polymerizations because most monomers and resins absorb negligibly at these wavelengths, yet industrial medium and high pressure mercury lamps exhibit strong emission bands in this region.

Photosensitization of diaryliodonium salts by anthracene occurs by a photoredox reaction in which an electron is transferred from an excited singlet or triplet state of the anthracene to the diaryliodonium initiator.[13-15,17]   The lifetimes of the anthracene singlet and triplet states are on the order of nanoseconds and microseconds respectively, and the bimolecular electron transfer reactions between the anthracene and the initiator are limited by the rate of diffusion of reactants, which in turn depends upon the system viscosity.   In this contribution, we have studied the effects of viscosity on the rate of the photosensitization reaction of diaryliodonium salts by anthracene.   Using steady-state fluorescence spectroscopy, we have characterized the photosensitization rate in propanol/glycerol solutions of varying viscosities.   The results were analyzed using numerical solutions of the photophysical kinetic equations in conjunction with the mathematical relationships provided by the Smoluchowski[16] theory for the rate constants of the diffusion-controlled bimolecular reactions.

Materials and Methods

Materials.   The 1-propanol, glycerol and anthracene were obtained from Aldrich Chemical Company and were used as received.   As in a previous study,[17] a commercially available bis-(4-dodecylphenyl) iodonium hexafluoroantimonate salt (UV9310C; GE Silicones) was used as the initiator. In this initiator, various dodecyl isomers have been attached to the phenyl rings of the diphenyliodonium salt to impart solubility in the monomer and do not effect the reactivity of the initiator.[1] All studies were performed with an order of magnitude excess of initiator relative to anthracene.

Fluorescence Measurements. The fluorescence experiments were performed in the *LASER Laboratory* at Michigan State University. The excitation source was a Coherent Innova 200 Argon ion laser operating at 363.8 nm. A Newport 845HP01 digital shutter system was opened with an electronic pulse from the detector controller to ensure the simultaneous start of illumination and fluorescence data acquisition. Unfocused laser radiation of 15 mW in a 3 mm diameter beam was directed upon the quartz capillary tube (1 mm i.d., 2 mm o.d., 25 mm long) placed perpendicular to the beam direction. The fluorescence was collected at a right angle from the incident beam and right angle from the longitudinal axis of the capillary tube. A Spex 1877 Triplemate spectrometer was used to analyze the fluorescence signal. The spectrometer had two stages: a subtractive dispersion filter stage and a spectrograph stage. A gated EG&G Princeton Applied Research Model 1530 C/CUV CCD detector, cooled to -120°C to minimize dark charge levels, was interfaced with an EG&G Princeton Applied Research Model 1463 OMA 4000 detector controller. The OMA 4000 software was used to analyze the results.

All steady state fluorescence experiments were conducted with the sample placed in a thermostated cell with temperature maintained at 30°C. The concentrations of anthracene and initiator used were 0.000505 and 0.00608 moles per liter, respectively. The relative quantities of solvents (n-propanol and glycerol) were adjusted from 0 to 100% to achieve solutions of different viscosities, while maintaining the same molar concentration of the reactive solutes.

Results and Discussion

In Situ Fluorescence Studies. Figure 1 contains a representative series of fluorescence spectra obtained *in situ* during the photosensitization of the diaryliodonium initiator by anthracene. The plot illustrates the decrease in the anthracene fluorescence as the photosensitization reaction proceeds (the spectra are relatively featureless due to the use of a low resolution grating). The initial slow response of the detector (which leads to an initial gradual increase in the fluorescence signal shown in the figure) may be attributed to an instrument artifact[15,17] and is not a consequence of the reaction mechanism. The actual fluorescence signal assumes a maximum value immediately after the shutter is opened and decreases monotonically in a nearly exponential decay. In our analysis, we will use the final 70 to 80% of the fluorescence data; therefore, the instrumental artifact is of no consequence, and important kinetic information can still be deduced.

The fluorescence decrease in Figure 1 can be attributed to the consumption of the anthracene photosensitizer during the photosensitization reaction. The photosensitization proceeds by an electron transfer reaction from the anthracene to the initiator, resulting in loss of aromaticity of the of the central ring.[17] Therefore, the photosensitization reaction leads to a disruption in the π electron structure of the anthracene, and the resulting molecule does not absorb at 364 nm (nor fluoresce in the 420 - 440 nm region). Hence, the steady-state fluorescence measurements allow the anthracene concentration to be monitored *in situ* while the photosensitization reaction takes place.

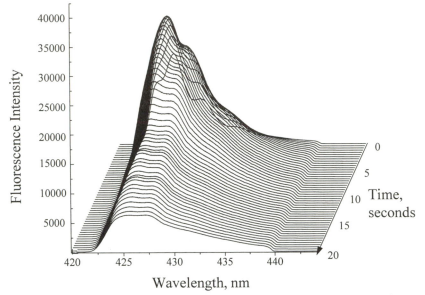

Figure 1.   Representative anthracene fluorescence profile during the photosensitization of diaryliodonium salts in pure propanol at 30°C.

A plot of the anthracene fluorescence intensity at 425 nm as a function of the
reaction time is shown in Figure 2.   Again, this figure exhibits the effect of the
instrumental artifact in the initial fluorescence data; however, examination of the
final 90% of the profile reveals that the anthracene concentration profile closely
follows a first order exponential decay.   Although the photosensitization reaction is
bimolecular, the anthracene concentration follows a pseudo-first-order profile since
the initiator is present in excess (i.e. $r = k_2 C_I C_A = k_1 C_A$ where r, C_I and C_A represent
the reaction rate, initiator concentration and ground state anthracene concentration,
respectively).   Therefore, the experimental results can be efficiently summarized by
the effective pseudo first-order decay rate constant, k_1.   This effective first order rate
constant, k_1, may be converted to the effective second order rate constant, k_2, by
dividing by the initiator concentration, C_I.   Note that k_1 and k_2 are not fundamental
rate constants for elementary reactions because the elementary reactions involve
excited state anthracene rather than the ground state.   These effective rate constant
actually embody a complex series of fundamental photophysical steps which
populate the excited states from the ground state, as described later in this paper.
However, adoption of these effective rate constants allows us to conveniently and
efficiently characterize the effect of viscosity on the photosensitization rate.

A series of steady-state fluorescence experiments were performed in mixtures
of propanol and glycerol to investigate the effect of viscosity on the effective second
order photosensitization rate constant, k_2.   Figure 3 illustrates that the effective rate
constant decreases as the viscosity of the system is increased.   For example, as the
reaction solvent is changed from pure propanol to pure glycerol, the viscosity of the
system rises by three orders of magnitude, while the effective reaction rate
coefficient, k_2, decreases by approximately one order of magnitude.

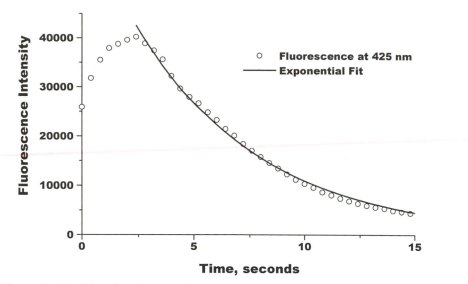

Figure 2. Profile of anthracene fluorescence at 425 nm obtained during photo-sensitization of diaryliodonium salt in pure propanol at 30°C.

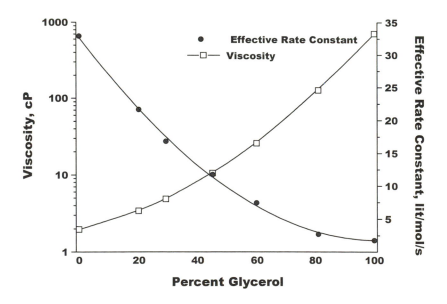

Figure 3. The effect of glycerol content on solution viscosity and the effective second order reaction rate constant, k_2.

Photophysical Kinetic Simulations. The transitions among the electronic energy levels of anthracene are shown schematically in Figure 4. As the figure illustrates, ground state anthracene may absorb a photon from the near-UV region of the spectrum to be promoted to the excited singlet state. The rate of this transition is governed by the first-order kinetic constant, k_{abs}, whose value depends upon the incident photon flux and the anthracene absorption cross section. Once promoted to the excited singlet state, the anthracene may take one of four reaction paths. It can revert back to the ground state with or without the simultaneous release of a fluorescence photon (governed by $k_{fluor} + k_{nonfluor}$). It can transfer an electron to a neighboring onium salt molecule to generate a cationic active center (governed by the product of $k_{senfluor}$ and the initiator concentration, $[I]$). Alternately, it may undergo intersystem crossing to form the triplet excited state (as described by the kinetic constant k_{isc}). Similarly, the triplet state anthracene may return to the ground state with or without simultaneous release of a phosphorescence photon ($k_{phos} + k_{nonphos}$) or may transfer an electron to the onium salt initiator (governed by the product of product of $k_{senphos}$ and $[I]$). Finally, the triplet state anthracene may donate an electron to an initiator molecule to form a reactive cationic species (the electron transfer reaction is thermodynamically feasible from both the singlet state and the triplet state of anthracene[17]).

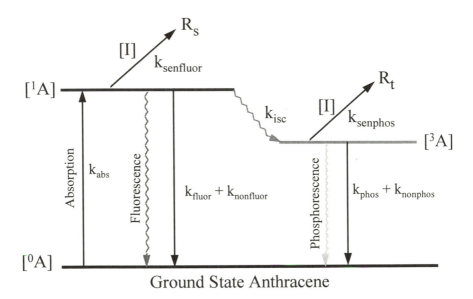

Figure 4. Electronic energy level diagram for anthracene illustrating the photophysical transitions (including reaction with the initiator from both the singlet and triplet states) and the associated kinetic constants.

Based upon the fundamental photophysical transitions outlined in Figure 4, a set of elementary chemical reactions may be written, as shown below. These reactions neglect side reactions such as photodimerization and oxidation due to their relative unimportance.[15] The effect of oxygen will be fully explored in a future publication.

$$A + h\nu \xleftarrow{k_{abs}} {}^1A \qquad \text{Eq. 1}$$

$${}^1A \xrightarrow{k_{fluor}} A + h\nu \qquad \text{Eq. 2}$$

$${}^1A \xrightarrow{k_{nonfluor}} A \qquad \text{Eq. 3}$$

$${}^1A \xrightarrow{k_{isc}} {}^3A \qquad \text{Eq. 4}$$

$${}^3A \xrightarrow{k_{phos}} A + h\nu \qquad \text{Eq. 5}$$

$${}^3A \xrightarrow{k_{nonphos}} A \qquad \text{Eq. 6}$$

$${}^1A + I \xrightarrow{k_{senfluor}} R_s \qquad \text{Eq. 7}$$

$${}^3A + I \xrightarrow{ksenphos} R_t \qquad \text{Eq. 8}$$

Here A, ${}^1A$, and ${}^3A$ represent anthracene in the ground state, the first excited singlet state and first excited triplet state, respectively. In addition, I represents the onium salt initiation, while R_s and R_t correspond to the reactive centers formed by reaction of the onium salt with the excited singlet and triplet state anthracene, respectively.

Based upon this set of elementary reactions, a series of coupled kinetic differential equations may be derived by taking material balances over the various reaction species, as shown below:

$$\frac{d[A]}{dt} = k_{abs}([{}^1A] - [A]) + (k_{fluor} + k_{nonfluor})[{}^1A] + (k_{phos} + k_{nonphos})[{}^3A] \qquad \text{Eq. 9}$$

$$\frac{d[{}^1A]}{dt} = k_{abs}([A] - [{}^1A]) - (k_{fluor} + k_{nonfluor})[{}^1A] - k_{isc}[{}^1A] - k_{senfluor}[{}^1A][I] \qquad \text{Eq. 10}$$

$$\frac{d[{}^3A]}{dt} = k_{isc}[{}^1A] - (k_{phos} + k_{nonphos})[{}^3A] - k_{senphos}[{}^3A][I] \qquad \text{Eq. 11}$$

$$\frac{d[I]}{dt} = -k_{senfluor}[{}^1A][I] - k_{senphos}[{}^3A][I] \qquad \text{Eq. 12}$$

$$\frac{d[R_s]}{dt} = k_{senfluor}[{}^1A][I] \qquad \text{Eq. 13}$$

$$\frac{d[R_t]}{dt} = k_{senphos}[{}^3A][I] \qquad \text{Eq. 14}$$

Solution of this coupled set of differential equations allows the concentrations of each of the anthracene electronic states to be determined as a function of time. In a previous publication, Nelson et al.[17] used this approach to investigate the relative importance of electron transfer from the singlet and triplet states of anthracene. In this contribution, we will use these simulations to predict profiles of the anthracene ground state as a function of time so that the simulation results may be compared with the steady-state fluorescence results presented above.

The kinetic parameters for this set of differential equations were either obtained from literature or were calculated based upon the experimental conditions. For example, based upon the experimental photon flux and extinction coefficient of anthracene at 365 nm, the value of k_{abs} was calculated to be 5.98 s^{-1} for our experiments. Based upon fluorescence lifetime experiments, the sum of the rate constants k_{fluor} and $k_{nonfluor}$ was determined by Nelson et al.[17] to be equal to 1.38 x 10^8 s^{-1}. Similar experiments with phosphorescence lifetime experiments have shown the sum of k_{phos} and $k_{nonphos}$ to be 1.0 x 10^5 s^{-1}.[17] Fluorescence and phosphorescence lifetime experiments conducted in the presence of initiator allowed the values of $k_{senfluor}$ and $k_{senphos}$ to be evaluated as 8.20 x 10^7 lit mol^{-1} s^{-1} and 2.50 x 10^6 lit mol^{-1} s^{-1} respectively.[17] Finally, the kinetic constant for intersystem crossing from the singlet to the triplet state for anthracene in solution has been reported to be between 7.38 x 10^7 to 1.12 x 10^8 s^{-1}.[18,19] An average of the reported values (0.929 x 10^8 s^{-1}) was utilized in our calculations.

The aforementioned values of the elementary kinetic constants were substituted into Equations 9 through 14, and the set of kinetic differential equations was solved simultaneously using the Rosenbrock method for stiff systems.[20] In agreement with the general experimental trend, the simulations predict that the anthracene concentration decreases exponentially with time (see Figure 5). Therefore, as discussed above, the overall rate of photosensitization may be efficiently characterized by the effective rate constants for photosensitization, k_1 or k_2 (again the rate of photosensitization may be written as r = $k_2 C_I C_A = k_1 C_A$ where k_1 is the effective first order rate constant, and k_2 is the effective second order rate constant). The kinetic simulations shown in Figure 5 may be used to predict the effect of the system viscosity on the effective photosensitization rate constant if the rate constants for the diffusion-controlled elementary reactions are modified to account for the medium viscosity. While the unimolecular reactions described by Equations 1 through 6 are relatively insensitive to viscosity, the bimolecular electron transfer reactions are known to be diffusion-controlled.[17] Therefore, the elementary rate constants $k_{senfluor}$ and $k_{senphos}$ must be modified to account for changes in the viscosity of the reaction medium.

The Smoluchowski theory for diffusion-controlled reactions, when combined with the Stokes-Einstein equation for the diffusion coefficient, predicts that the rate constant for a diffusion-controlled reaction will be inversely proportional to the solution viscosity.[16] Therefore, the literature values for the bimolecular electron transfer reactions (measured for a solution viscosity of η_1) were adjusted by multiplying by the factor η_1/η_2 to obtain the adjusted value of the kinetic constant appropriate for a solution with a viscosity equal to η_2. A series of kinetic simulations were performed by solving Equations 9 through 14 using $k_{senfluor}$ and $k_{senphos}$ which were adjusted in this manner (all the other kinetic constants were unchanged) to account for system viscosities varying over three orders of magnitude from 1 to 1000 cP. Figure 5 contains the simulation results in the form of anthracene concentration profiles which illustrate the effect of viscosity. The figure illustrates that at low viscosities, the decay rate constant is relatively high (indicating that the anthracene concentration exhibits a relatively rapid exponential decay); while at higher solution viscosities, the exponential decay rate is much slower. For example, the figure shows that at a relatively low viscosity of 0.9 cP, the simulations predict that anthracene concentration has decreased to zero in less than ten seconds; while at a viscosity of 675 cP, the simulations predict that the anthracene concentration is essentially unchanged from its initial value even after twenty seconds.

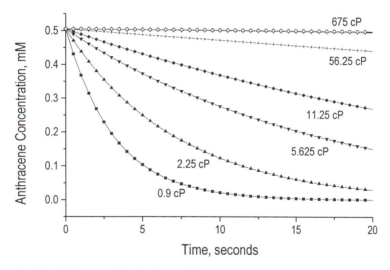

Figure 5. Simulation results for the anthracene concentration as a function of time during the photosensitization reaction at different viscosities.

Comparison of the Experimental and Simulation Results. The preceding discussion has shown that both the experimental anthracene fluorescence profiles and the simulated anthracene concentration profiles decrease in a manner which closely follows an exponential decay. Therefore, the most convenient way to compare the simulation results to the experimental data is to define an effective overall photosensitization rate constant, k_1 or k_2, as described above. Adoption of this lumped-parameter effective kinetic constant allows us to conveniently and efficiently compare the experimental data to the simulation results by contrasting the rate constant obtained from the steady-state fluorescence decay with the value obtained from the simulated decrease in the anthracene concentration.

Figure 6 shows a comparison of the experimental data with the simulation results for the effect of viscosity on the effective second order rate constant for photosensitization, k_2. In this figure, the simulation results based upon the Smoluchowski-Stokes-Einstein model are indicated by the solid line. This curve was generated by solving Equations 9 through 14 simultaneously for a variety of viscosities in which the elementary rate constants for the diffusion-controlled bimolecular reactions, $k_{senfluor}$ and $k_{senphos}$, were scaled with the inverse viscosity. Comparison of this curve with the experimental values for the effective rate constant (represented in the figure by squares) reveals that while the simulations describe the general experimentally observed trend, there is no quantitative agreement. For example, both the experimental and the simulation results indicate that the effective photosensitization rate constant falls an order of magnitude as the viscosity is increased from 1 to 1000 cP; however, the simulation predicts a more drastic reduction than observed experimentally. Therefore, the effective rate constant is less viscosity-dependent than the Smoluchowski-Stokes-Einstein model predicts.

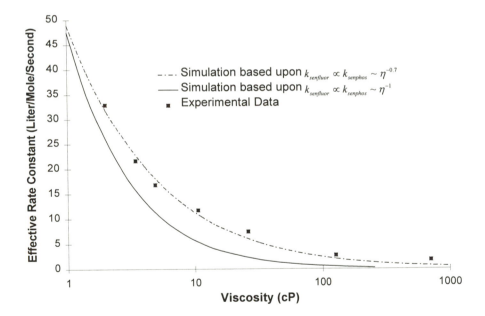

Figure 6. Comparison of the experimental data with simulation results for the effective photosensitization rate constant, k_2, as a function of viscosity.

The discrepancies between the experimental data and the behavior predicted using the Smoluchowski-Stokes-Einstein model for $k_{senfluor}$ and $k_{senphos}$ likely arise from the inadequacies of the simple Smoluchowski-Stokes-Einstein analysis for application to the anthracene/diaryliodonium salt molecular system. For example, the Smoluchowski analysis assumes that the reacting molecules are spherical in shape and of approximately equal size. In reality, the anthracene molecule has fused aromatic ring architecture and a rigid, planar shape whereas the initiator molecule, with its linear chains of 20 carbon atoms and two phenyl rings, has a long and narrow molecular structure. In addition, the Stokes-Einstein equation is known to provide only an approximate estimate of the viscosity dependence of the diffusion coefficient, especially when applied on the molecular scale.[16]

If the viscosity dependence of the elementary bimolecular kinetic constant is modified from the simple inverse relationship suggested by the Smoluchowski-Stokes-Einstein analysis, a much better quantitative agreement between the simulation results and the experimental data can be obtained. For example, the dotted curve in Figure 6 illustrates the simulation results obtained by scaling the elementary rate constants $k_{senfluor}$ and $k_{senphos}$ with $\eta^{-0.7}$ (rather than the η^{-1} as suggested by the Smoluchowski-Stokes-Einstein analysis). Again this curve was obtained by solving the set of kinetic differential equations (Equations 9 through 14) for a variety of viscosities with the two bimolecular rate constants scaled with $\eta^{-0.7}$ and all other kinetic constants unchanged. Figure 6 illustrates that this simple modification leads to excellent agreement between the experimental data and the simulation results. Therefore, this curve provides a semi-empirical correlation which may be used to estimate the effective photosensitization rate constant for viscosities ranging from 1 to 1000 cP.

Conclusions

The experimental and simulation results presented here indicate that the system viscosity has an important effect on the overall rate of the photosensitization of diaryliodonium salts by anthracene. These studies reveal that as the viscosity of the solvent is increased from 1 to 1000 cP, the overall rate of the photosensitization reaction decreases by an order of magnitude. This decrease in reaction rate is qualitatively explained using the Smoluchowski-Stokes-Einstein model for the rate constants of the bimolecular, diffusion-controlled elementary reactions in the numerical solution of the kinetic photophysical equations. A more quantitative fit between the experimental data and the simulation results was obtained by scaling the bimolecular rate constants by $\eta^{-0.7}$ rather than the η^{-1} as suggested by the Smoluchowski-Stokes-Einstein analysis. These simulation results provide a semi-empirical correlation which may be used to estimate the effective photosensitization rate constant for viscosities ranging from 1 to 1000 cP.

The results presented here suggest that the large change in viscosity which occurs during a photopolymerization reaction will have a profound effect on the rate of active center production by photosensitization. The lumped-parameter effective photosensitization rate constant will assume the largest value at the beginning of the reaction when the viscosity is the lowest. As the reaction proceeds and the viscosity of the mixture increases due to polymerization, the rate of photosensitization will decrease substantially. It is therefore expected that the rate of initiation will become extremely low after a certain degree of polymerization, and that any further reaction will be almost exclusively from propagation of existing active centers.

Acknowledgments

The experimental studies reported in this paper were performed in the Michigan State University *LASER* Laboratory. In addition, this research was supported by the Michigan State University Crop and Food Bioprocessing Center.

Literature Cited

1.   Crivello, J. V.; Narayan, R. *Chem. Mater.* **1992,** *4,* 692.
2.   Pappas, S. P.; *UV Curing, Sci. and Tech.*, Technology Marketing Corp., Norwalk, CT, 1985, Vol. 2.
3.   Crivello, J. V.; *Adv. Polym. Sci.* **1984,** *62,* 1.
4.   Reiser, A. *Photoreactive Polymers*, Wiley, New York, NY, 1989.
5.   Lapin, S. P. In *Radiation Curing of Polymeric Materials;* Hoyle, C. E.; Kinstle, J. F., Eds., ACS Symposium Series, 417, ACS: Washington DC, 1989; pp 363.
6.   Pappas, S. P. *Prog. Org. Coatings*, **1985,** *13,* 35.
7.   Crivello, J. V.; Lam, J. H. W. In *Epoxy Resin Chemistry;* Bauer, R. S., Ed.; ACS Symposium Series, 114, ACS, Washington DC, 1979, pp 1.
8.   Crivello, J. V.; Parfitt, G. D.; Patsis, A. V., Eds., Organic Coatings, Science and Technology; Dekker, New York, NY, 1983, Vol. 5; pp 35.
9.   Watt, W. R. In *Epoxy Resin Chemistry*, R.S. Bauer, Ed., ACS Symposium Series, 114, ACS, Washington DC, 1979, pp 17.
10.   Lohse, F.; Zweifel, H. *Adv. Polym. Sci.* **1986,** *78,* 61.
11.   Crivello, J. V.; Conlon, D. A. *J. Polym. Sci. Part A: Polym. Chem.* **1983,** *21,* 1785.
12.   Crivello, J. V.; Lee, J. L. *J. Polym. Sci. Part A: Polym. Chem.* **1989,** *27,* 3951.
13   Sundell, P.; Jonsson, S.; Hult, A. *J. Polym. Sci., Part A: Polym. Chem.* **1991,** *29,* 1525.
14.   Manivannan, G.; Fouassier, J. P. *J. Polym. Sci., Part A: Polym. Chem.* **1991,** *29,* 1113.
15.   Nelson, E. W.; Carter, T. P.; Scranton, A. B. *Macromolecules* **1994,** *27,* 1013.
16.   Bamford, C. H.; Tipper, C. F. H.; Compton, R. G., Eds., Comprehensive Chemical Kinetics; Elsevier, New York, NY, 1985; Vol. 25.
17.   Nelson, E. W.; Carter, T. P.; Scranton, A. B. *J. Poly. Sci., A: Polym. Chem.* **1995,** *33,* 247.
18.   Kavarnos, G. J. In *Photoinduced Electron Transfer*, Dewar, M. J. S.; Dunitz, J. D.; Hafner, K.; Ito, S; Lehn, J. M.; Niedenzu, K.; Raymond, K. N.; Rees, C. W.; Vogtle, F., Eds., Topics in Current Chemistry, 156, Springer, New York, NY, 1990, 27.
19.   Turro, N. J. *Molecular Photochemistry*, W. A. Benjamin, New York, NY, 1965, 86.
20.   Borrelli, R. L.; Coleman, C. S. *Differential Equations - A Modeling Approach*, Prentice-Hall, Englewood Cliffs, NJ, 1987.

Chapter 9

Alkenyloxystyrene Monomers for High-Temperature Adhesives and Sealants

John Woods[1], Maria Masterson[1], Ciaran McArdle[2], and Joe Burke[2]

[1]Loctite Corporation, 1001 Trout Brook Crossing, Rocky Hill, CT 06067
[2]Loctite Ireland, Whitestown Business Park, Tallaght, Dublin 24, Ireland

Cationically curable compositions derived from 4-allyloxy-styrene monomers have been found to be useful for the production of UV and heat curable adhesives and sealants having outstanding thermal resistance properties. The monomer compositions are cured in two stages: firstly, by a UV initiated cationic polymerization, and secondly, by a thermally promoted Claisen Rearrangement, to yield an activated phenolic polymer. On formation, the Claisen product undergoes a rapid intermolecular electrophilic addition of the rearranged alkenyl group, to produce a highly crosslinked and chemically resistant polymer having a decomposition temperature in excess of 400°C. The chemical and physical properties of the new adhesive materials are described in this review.

Polyimides are a well known class of high performance polymers with many important commercial applications in the area of adhesives, coatings and composite materials. They are generally associated with good thermal stability and mechanical properties, but are difficult to process due to the insolubility and poor melt processability of the polymer. In most cases, they are applied as dilute solutions of poly(amic acids) (imide precursor) in organic solvent, dried and cured at elevated temperatures. However, a disadvantage with this method is that the poly(amic acid) solutions are hydrolytically unstable and deteriorate on storage, leading to a corresponding loss in physical properties of the cured polyimide *(1)*. In addition, large quantities of organic solvent have to be removed and/or recovered from the processing of a polyimide, which gives rise to environmental concerns. Consequently, there exists a need for an easily processable and thermally resistant, solventless composition, that exhibits good long term storage in the uncured state. Our research on thermally resistant adhesive compositions has focused on a relatively new and little studied class of cationically polymerizable substituted styrene monomers.

Monomers and reactive oligomers containing multiple 4-alkoxystyrene (styryloxy) or *para* vinylphenol ether functional groups are useful materials for the production of photopolymerizable adhesives and coatings *(2-4)*. They undergo photo-initiated cationic polymerization at rates that are comparable to those of vinyl ethers and provide cured materials with superior properties compared to alkyl or aryl vinyl ether polymers *(5)*. The development of commercial products derived from these materials has been limited, however, owing in part to the relatively high cost of their production (usually multi-step synthesis) and in part to the poor solubility of commercially available onium salt cationic photocatalysts *(4)*. In contrast, low molar mass mono-functional alkoxystyrene monomers are excellent solvents for the photocatalysts, but generally provide only low molecular polymers with weak mechanical strength and poor chemical resistance.

We have recently discovered, however, that certain 4-alkenyloxystyrene monomers have good compatibility with cationic photoinitiators, polymerize rapidly under UV light and, following a thermal post-cure, provide crosslinked products with outstanding thermal resistance and high glass transition temperatures *(6,7)*. In addition, these materials have low viscosity, show good adhesive properties and do not contain organic solvents. They have been found useful for the assembly of a variety of electronic and optical components. In this chapter, we will outline the synthesis, polymerization and properties of photocured films derived from 4-allyloxystyrene and blends of this monomer with a divinyl ether.

Experimental

4-Allyloxystyrene was prepared from 4-allyloxybenzaldehyde and methyltriphenylphosphonium bromide under Wittig reaction conditions as previously described *(2)*. Photocurable compositions were prepared by dissolving 1.5 weight % of cationic photoinitiator Cyracure UVI 6974 (50% solution of mixed triarylsulfonium hexafluoroantiminoate salts in propylene carbonate; supplied by Union Carbide) in the monomer or monomer blend. Triethyleneglycol divinylether (DVE-3, supplied by GAF) was employed as the crosslinking agent in samples prepared for mechanical and photo DSC analysis. Polymer films for IR analysis were cured on glass slides under controlled exposure to a high pressure short arc mercury lamp illuminator, equipped with a 290-390 nm dichroic mirror and quartz collimating lens (supplied by Oriel Corporation). The incident light intensity was adjusted to 30 mW/cm^2. IR analyses were carried out on a Mattson Genesis FTIR, at a resolution of 4 cm^{-1}.
Dynamic mechanical analysis (DMA) was performed on a Polymer Labs Analyzer at a frequency of 1 Hz and TGA data was obtained on a TA 951 thermogravimetric analyzer.

Molecular weight measurements were determined by size exclusion chromatography on a high pressure liquid chromatograph equipped with a differential refractometer. A Waters "Styragel HR" 5μ column set (10^6; 10^4; 500 A) was employed and calibrated with PS standards. THF was used as solvent at a flow rate of 1 ml/min.

Adhesion tests were performed according to ASTM D2095, using cylindrical steel pins bonded to glass plate and were cured by UV exposure, through the glass plate, to a UVALOC 1000 lamp (supplied by Loctite Deutschland GmbH).

Results and Discussion

Monomer Synthesis. 4-Allyloxystyrene was prepared by the Wittig reaction of 4-allyloxybenzaldehyde and methyltriphenylphosphonium bromide, under basic conditions. The allyloxybenzaldehyde was prepared, in turn, by the alkylation of 4-hydroxybenzaldehyde with allyl bromide. This method, which provides high purity monomer in high overall yield, is outlined in Scheme 1 and has been previously described *(2)*. Alternatively, the monomer may be prepared by the direct alkylation of *p*-vinylphenol with allyl bromide *(8,9)*, although this method is less convenient due to the difficulties in synthesizing and storing the highly reactive vinyl phenol *(10)*.

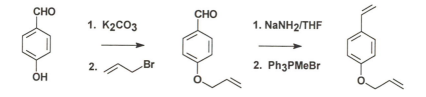

Scheme 1.

Several related alkenyloxystyrene monomers, including 4-methallyl-oxystyrene (1), were also prepared by a similar procedure. Related monomers, 4-(1'-propenyloxy)styrene (2) and 4-(2'-methyl-1'-propenyloxy)styrene (3) were prepared by the base catalyzed isomerization of allyl and methallyloxy styrenes respectively *(6)*. The propenyloxystyrene derivatives were found to be particularly useful as low viscosity crosslinking agents of allyoxystyrene monomers, since the propenyloxy group exhibits a much higher cationic reactivity compared to the allyloxy function. Several commercially available divinylether materials were also found to be suitable as crosslinking agents.

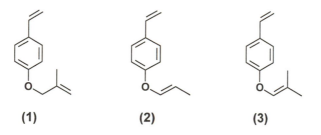

(1)   (2)   (3)

UV Photopolymerization (A-Stage Cure). 4-Allyloxystyrene is a colorless liquid having a viscosity < 10 mPa.s at 25°C. The IR spectrum of the monomer shows peaks at 902 and 926 cm^{-1} due to the hydrogen out of plane deformations of the styrene and allyl alkenyl groups respectively (Figure 1a). Two additional characteristic absorption bands due to the double bond stretching vibration of the styrene and allyl groups are observed at 1627 and 1647 cm^{-1} respectively *(11)*. On exposure to UV light for 10 seconds, a thin film sample of the monomer, containing 0.75% by weight of cationic photoinitiator, polymerized to give a dry, cured film that was soluble in acetone. IR analysis of the cured film showed that the absorbance peak of the styrene double bond at 902 cm^{-1} had disappeared, whereas the peak due to the allyl group at 926 cm^{-1} was still present (Figure 1b). Similarly, the styrene absorbance at 1627 cm^{-1} had disappeared, whereas the peak due to the allyl group at 1647 cm^{-1} was still present. It may be concluded, therefore, that the structure of this material is that of linear product, poly(4-allyloxystyrene), as indicated in Scheme 2. SEC analysis showed the polymer to have an number average molecular weight of 24,000 and a molecular weight distribution of 2.4. This result is in contrast to previous work on the bulk cationic polymerization of 4-allyloxystyrene, which found that treatment of the monomer with aluminum chloride at 20°C yielded an insoluble crosslinked polymer *(8)*. However, solution polymerization of the monomer in dichloromethane at -78°C, using boron trifluoride etherate as initiator, was found to give a linear soluble polymer, having a similar structure to the product we obtained on UV exposure *(9)*.

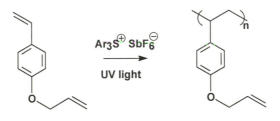

$$Ar_3S^{\oplus}\ SbF_6^{\ominus}$$
$$\xrightarrow{\hspace{2cm}}$$
UV light

Scheme 2.

Crosslinked co-polymers of 4-allyloxystyrene can be obtained by the addition of small amounts of divinylethers, di-functional alkoxystyrene monomers or propenyloxystyrene monomers, such as (2) or (3), in the cationically active composition. The polymers obtained from these mixtures by cationic polymerization are insoluble in organic solvents and generally exhibit good mechanical and adhesive properties.

Under free radical conditions, we found that the bulk polymerization of 4-allyloxystyrene gave an insoluble crosslinked polymer with AIBN. Similar results were previously reported for the polymerization with benzoyl peroxide *(8)*. The cationic polymerization of *p*-alkoxystyrene monomers have been shown to proceed at rates that are comparable to vinyl ethers *(12,13)*. As expected from these studies, we found that alkenyloxystyrene monomers also have a high degree of cationic reac-

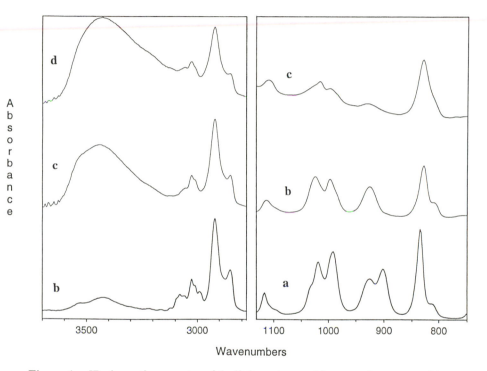

Figure 1. IR absorption spectra of 4-allyloxystyrene (a) uncured monomer (b) after 10 s UV exposure with cationic photoinitiator (c) after UV exposure and heating at 200°C for 1 hour and (d) after an additional 16 hours heating at 200°C.

tivity. Photo DSC analysis (Figure 2) confirmed that the UV light response of 4-allyloxystyrene and compositions derived from the monomer, in the presence of a catalytic amount of a triarylsulfonium salt photoinitiator (0.75% by weight), have almost identical photo responses to that of the commercially available divinyl ether monomer, DVE-3. In addition, we have found that solutions of commercially available cationic photocatalysts in the monomer and monomer blends are optically clear.

Thermal Crosslinking (B-Stage Cure). On heating, the UV cured polymers and copolymers undergo a number of structural changes. The materials develop an intense deep red color and become brittle and insoluble in organic solvent. During this process, the allyl group absorbances at 926 and $1647 cm^{-1}$ in the IR spectrum disappear (Figure 1c shows the spectrum in the region of 900 cm^{-1}) and an intense hydroxyl group band centered at 3445 cm^{-1} is formed, as indicated in Figures 1b-d. These changes are consistent with the initial formation of poly(3-allyl-4-hydroxystyrene) by a Claisen Rearrangement reaction, followed by an electrophilic addition of the allyl or isomerized allylic group to the newly formed phenol to give a crosslinked phenolic type resin, as outlined in Scheme 3.

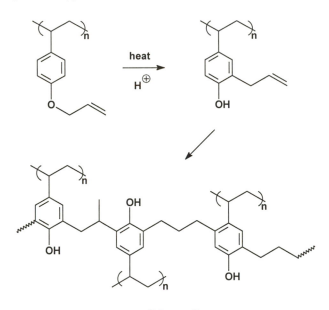

Scheme 3

The formation of phenolic polymers by Claisen Rearrangement of poly(4-allyloxystyrenes) under acid catalyzed thermolysis conditions has previously been reported in connection with the development high resolution photoresists *(14,15)*. This work was primarily focused on the production of soluble phenolic polymers that could be imaged on the basis of differential dissolution. In this regard, allyloxystyrene polymers bearing alkyl substituents at the α-position to the ether oxygen atom

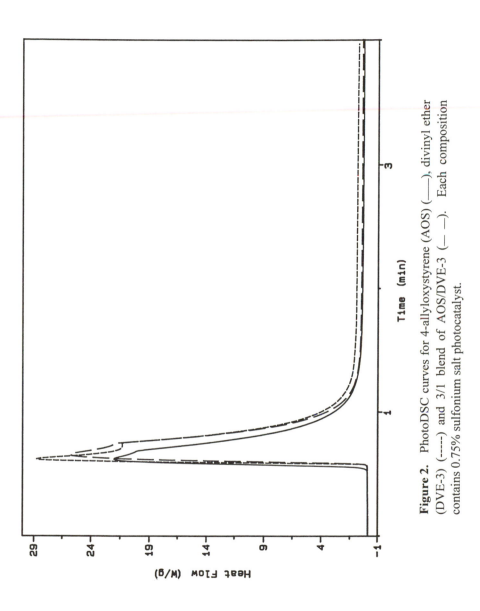

Figure 2. PhotoDSC curves for 4-allyloxystyrene (AOS) (——), divinyl ether (DVE-3) (----) and 3/1 blend of AOS/DVE-3 (— —). Each composition contains 0.75% sulfonium salt photocatalyst.

were quantitatively converted to the corresponding soluble phenolic polymers by a combination of elimination and Claisen Rearrangement reactions as indicated in Scheme 4. In contrast, allyloxystyrene polymers that can not eliminate a conjugate diene, such as poly(4-allyloxystyrene), were found to be less attractive for imaging applications due to swelling during development *(15)*. This swelling is likely due to the formation of crosslinked polymer according to the mechanism outlined above.

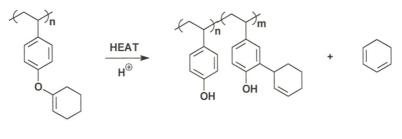

Scheme 4

The post-irradiation Claisen Rearrangement may be efficiently performed in the neat polymer films at temperatures of 140-320°C. Both the rate and extent of phenol formation increase as the temperature is raised. The extent of allyl group consumption was estimated by measuring the absorbance ratio of the bands at 926 and 828 cm^{-1} (as an internal standard) before and after heating at 200°C. After 1 hour, the conversion was found to be in excess of 80%, but further conversion occurred only relatively slowly, as shown in Figure 3. The dramatic reduction in post-curing rate, observed after the first hour can be attributed to the onset of vitrification (see DMA data below) at which stage there is insufficient diffusion of the polymer bound reactive carbenium ion species to permit additional crosslinking. It can be predicted, however, that performing the B-stage reaction at a temperature in excess of 200°C would enhance both the rate and extent of post-curing.

It appears from the IR data, that the allylic group is consumed as the Claisen rearrangement occurs and not in two distinct reaction steps. This suggests that the Claisen Rearrangement and not the ene/phenol addition reaction, is the rate limiting step in the overall B-stage process. The high reactivity of the allyl addition reaction can be attributed to the formation of the phenol, which strongly activates the *ortho* site for electrophilic addition and to the presence of Lewis acid residues produced during the decomposition of the photoinitiator, which function as the catalyst for the electrophilic addition reaction.

Thermal and Dynamic Mechanical Analyses. The thermal and mechanical properties of the new photocured films were evaluated by DMA and TGA. The initial DMA scan of the photocured polymer (prepared from a 3:1 blend of 4-allyloxystyrene: DVE-3) to ensure crosslinking on UV exposure shows two transitions at 62°C and 232°C, as shown in Figure 4, scan a. The low temperature transition is associated with a reduction in the storage modulus (G') and can therefore be attributed to the glass transition (T_g) of the polymer. The high temperature process is initially associated with a small drop in modulus, but as the temperature is increased beyond the peak value, the modulus begins to increase again. This process can therefore be attributed to the onset of the secondary post-curing reaction as outlined above. The remaining scans, depicted in Figures 4b-d, were obtained by cooling the sample to ambient temperature and then re-scanning. The rescans show that both of the original loss processes have disappeared and that the modulus shows no change over the temperature range ambient to 300°C. This indicates that a highly crosslinked and hard material, having a T_g higher than 300°C, has been formed as a result of the secondary post-curing reaction.

The new materials have outstanding thermal resistance properties. The onset of decomposition of the post-cured resin, as determined by dynamic TGA, under nitrogen, was 430°C (see Figure 5). In air, the corresponding value was 423°C indicating good oxidative stability. Isothermal TGA were performed at 300°C and showed no weight loss after 8 hours (Figure 6). A similar result was recorded at 325°C and a significant weight loss was not observed under these conditions until the temperature was increased to 350°C.

In contrast, the polymer obtained by the cationic photopolymerization of divinyl ether, DVE-3, showed significant weight losses at temperatures in excess of 200°C. In dynamic TGA tests, the onset of polymer decomposition was found to be 370°C for this material.

Dielectric Properties. The dielectric properties of the fully cured poly(4-allyloxystyrene) and the 3/1 blend of poly(4-allyloxystyrene)/DVE-3 (divinylether) composition were investigated over the frequency range 1 MHz to 1 GHz and the data, with respect to the pure polymer, measured at 23°C, is presented in Table I. Similar values were observed for the post-cured poly(4-allyloxystyrene)/DVE-3 co-polymer. The dielectric constant values obtained are similar to those reported for electronic grade polyimides although the dissipation factors (tan δ values) are slightly higher *(16)*. The dielectric constant values were found to increase on exposure to high humidity, but showed good recovery on subsequent standing under ambient conditions.

Table I. Dielectric properties of UV and heat cured poly(4-allyloxystyrene)

Frequency (MHz)	ε'	ε''	tan δ
1	3.4	0.059	0.017
10	3.3	0.057	0.017
100	3.2	0.054	0.017
1000	3.1	0.052	0.017

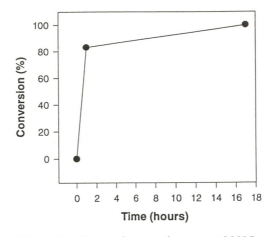

Figure 3.   Post-curing reaction rate at 200°C.

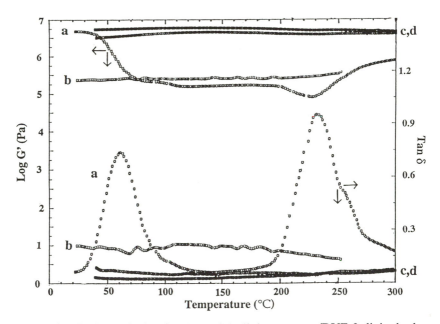

Figure 4.   DMA analysis of UV cured 4-allyloxystyrene/DVE-3 divinyl ether blend (3:1) at 1 Hz and 5°C/min., (a) initial scan; (b,c,d) re-scans of the same sample.

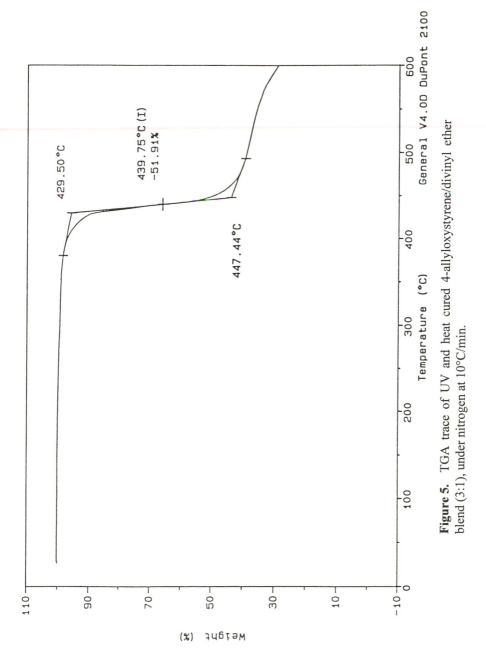

Figure 5. TGA trace of UV and heat cured 4-allyloxystyrene/divinyl ether blend (3:1), under nitrogen at 10°C/min.

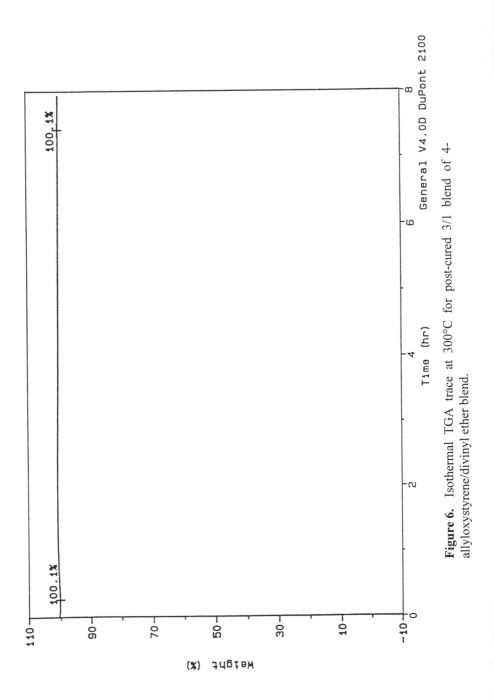

Figure 6. Isothermal TGA trace at 300°C for post-cured 3/1 blend of 4-allyloxystyrene/divinyl ether blend.

Adhesive Properties. The adhesive strength properties of a 3:1 blend of 4-allyloxy-styrene/DVE-3 (divinyl ether) composition containing 0.75 % of the sulfonium salt photocatalyst was examined by measuring the tensile breaking strength of cylindrical steel pins bonded to glass plates. The measurements were made according to the standard test method, ASTM D2095, and the test specimens were cured by exposure to UV light through the glass substrate. The light intensity, measured at 365 nm was 100 mW/cm^2. The results obtained for different UV exposure times are presented in Table II. and are the average values obtained for 6 bonded assemblies per test. The data show that high strength development occurs following very short irradiation times. At the optimum adhesive strength value (12-14 MPa), the cationically cured allyloxystyrene product has approximately 80% of the strength that is usually associated with a UV curable glass bonding acrylic adhesive (15-18 MPa) tested under the same conditions. The adhesive strength of the fully cured product after post-curing has yet to be determined. However, qualitative assessments indicate that adhesive strength on non-flexible substrates is adequate for many assembly operations.

Table II. Adhesive test results for 3:1 blend of 4-allyloxstyrene/DVE-3

UV Exposure Time (secs)	Bond Strength (MPa)
5	10.8
20	13.5
40	12.4
60	10.1

Conclusions

Alkenyloxystyrene monomers such as 4-allyloxystyrene are useful components of photocured cationically polymerizable compositions. Used alone or in combination with divinyl ethers they provide low viscosity formulations, which are excellent solvents for commercial onium salt photoinitiators. Photocuring rates are comparable to vinyl ether monomers and the initially photocured alkenyloxystyrene polymers may be further heat processed to yield crosslinked phenolic type resins having outstanding thermal resistance properties. The new materials have good adhesive properties and are potentially useful where a combination of ease of processability and high performance is required.

Literature Cited

1. Perry, R. J.; Tunney, S. E.; Wilson, B. D. *Macromolecules* **1996**, *29*, 1014.
2. Woods, J; Rooney J. M.; Harris, S. *U.S. Patent 4,543,397*, **1985**, (Loctite Corp.).
3. Ericsson, J.; Hult, A. *Polym. Bull.* **1987**, *18*, 295.
4. Crivello, J. V.; Ramdas, A. *J Macromol. Sci., Pure Appl. Chem.* **1992**, *A29* (9), 753.

5. Crivello, J. V.; Suh, D-H. *J. Polym. Sci., Polym. Chem.* **1993**, *31* (7), 1847.
6. McArdle, C. B.; Burke, J.; Woods, J. *U. S. Patent 5,084,490*, **1992**, (Loctite Corp.).
7. McArdle, C. B.; Burke, J.; Woods, J. *U. S. Patent 5,141,970*, **1992**, (Loctite Corp.)
8. D'Alelio, G. F.; Hoffend, T. R. *J. Polym. Sci.: Part A-1* **1967**, *5*, 1245.
9. Kato, M.; Kamogawa, H. *J. Polym. Sci.: Part A-1* **1968**, *6*, 2993.
10. Coroson, B. B.; Heintzelman W. J.; Schwartzman, L. H.; Tiefenthal, H. E.; Lokken, R. J.; Nickels, J. E.; Atwood G. R; Pavlik F. J. *J. Org Chem.* **1958**, *23*, 544.
11. Bellamy, L. J. *The Infra-red Spectra of Complex Molecules;* J. Wiley & Sons: N.Y. **1954**; p 34.
12. Kanoh, N; Gotoh, A.; Higashimura, T.; Okamura, S. *Makromol. Chem.* **1963**, *63*, 115.
13. Cotrel, R; Sauvet, G; Vairon, J. P.; Sigwalt, P. *Macromolecules,* **1976**, *9* (6), 931.
14. Fréchet, J. M. J.; Kallman, N; Kryczka, B.; Eichler, E.; Houlihan, F. M.; Willson, C. G. *Polym. Bull.* **1988**, *20*, 427.
15. Fréchet, J. M. J.; Eichler, E.; Gauthier, S.; B. Kryczka, B.; Willson, C. G. in *The Effects of Radiation on High Technology Polymers*; Editors, Reichmanis, E.; O'Donnell, J. H.; ACS Symposium Series 381; Americal Chemical Society: Washington, DC, **1989**; p 155.
16. Schuckert, C. C.; G. B. Fox, G. B. in *New Technology in Electronic Packaging,* Editors, Livesay, B. R.; Nagarkar, M. D.; ASM International, Materials Park, OH, **1990**; pp 91-97.

Chapter 10

Maleate/Vinyl Ether UV-Cured Coatings: Effects of Composition on Curing and Properties

Gerry K. Noren

DSM Desotech, Inc., 1122 St. Charles Street, Elgin, IL 60120

The effect of the composition of the maleate polyesters and vinyl ethers used on UV-curing and film properties has been investigated. Linear unsaturated polyesters were prepared from maleic anhydride with succinnic anhydride and 1,5-pentane diol. The equivalent weight and molecular weight of the unsaturated polyesters was varied. Coating formulations containing these unsaturated polyesters, triethylene glycol divinyl ether and a free radical photoinitiator were crosslinked in the presence of UV light. A ratio of vinyl ether equivalents to maleate equivalents of 1:1 was preferred for optimal crosslinking. Photoinitiators 2-hydroxy-2-methyl propiophenone and diphenyl-(2,4,6-trimethylbenzoyl) phosphine oxide produced the fastest cure. Photoinitiator concentration had little effect on the degree of cure of the coating. Diethyl maleate and isobutyl vinyl ether were effective for reducing viscosity but reduced cure speed. A vinyl ether urethane oligomer was synthesized which enhanced flexibility and toughness of the coatings when substutited for triethylene glycol divinyl ether.

The development of commercial UV/EB curable coating systems based on non-acrylate chemistry is an objective of the UV/EB industry. Over the years, several non-acrylate UV curable systems have been reported in the literature. These include, cationic initiated systems based on either cycloaliphatic epoxies[1] or vinyl ethers[2-4] and free radical systems based on thiol/ene[5-7] or amine/ene[8,9] reactions, as well as, various hybrid systems. Our recent research has been directed toward the investigation of the photoinduced 1:1 alternating copolymerization reaction of vinyl ethers with maleate esters[10,11] as a crosslinking reaction for UV cured coatings. The objective of this research was to evaluate the effects of oligomer molecular weight and functionality, photoinitiator and diluent on film properties and coating properties.

Free radical alternating 1:1 copolymerization of donor-acceptor monomer systems has been known for quite some time. Linear copolymers from donor-acceptor systems have been formed by both thermal and photolytic initiation techniques.[12] Commercially available linear styrene/maleic anhydride copolymers and vinyl ether/maleic anhydride copolymers are good examples of this copolymerization reaction. An example of a thermosetting system is the peroxide/redox initiated crosslinking of the unsaturated polyester-styrene systems used in fiber glass composites.

Photoinduced copolymerization of donor-acceptor monomer pairs (Scheme 1) can be either self initiated by excitation of the charge transfer complex (charge transfer initiation) or by polymerization of the charge transfer complex/monomer equilibrium

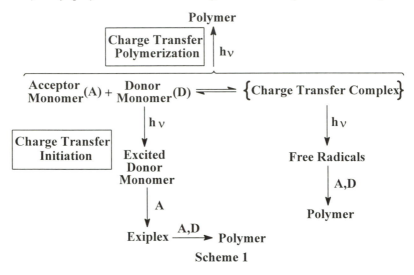

Scheme 1

mixture initiated by a photoinitiator (charge transfer polymerization).[12,13] In charge transfer initiation, polymerization occurred in strong donor/weak acceptor and weak donor/strong acceptor pairs while in charge transfer polymerization no relationship between polymerization and donor-acceptor strength was observed.[14]

Experimental

The chemicals used in syntheses were used as received from the supplier. The unsaturated polyesters were prepared by standard methods from the diacids and diols using butyl stannoic acid (Fascat 4100) as the catalyst. The water of condensation removed using a xylene azeotrope and after reaching the desired acid value the xylene was removed by distillation. The synthesis of unsaturated polyester A is representative.

Synthesis of Unsaturated Polyester A. The unsaturated polyester oligomer A was prepared in a 500 milliliter 4-neck flask fitted with a variable speed stirrer, a heating mantle, a gas inlet tube, a thermometer, a Dean-Starke trap and a condenser. The flask was charged with 147.0 g (1.5 mol) of maleic anhydride (Amoco), 195 g (1.88 mol) of 1,5-pentane diol (BASF), about 35 ml of xylene and 0.3 g of Fascat 4100. Then heated slowly under a nitrogen blanket to 175° C while the water of condensation was removed as a xylene azeotrope. The reaction was held until the acid value was less than 10.0 and the xylene was removed by distillation. The resultant resin had an acid value of 7.5, a viscosity of 3,340 mPa·s at room temperature and a theoretical equivalent weight of 210 g per maleate double bond.

Synthesis of a Urethane Vinyl Ether Oligomer.   A urethane vinyl ether oligomer was prepared by the procedure discussed by Lapin.[15] Reacting a hydroxyalkyl vinyl ether, an oligomeric diol and a diisocyanate gave and oligomer that had a Brookfield viscosity of 3.7 million mPa·s and a theoretical equivalent weight of 1000 g per vinyl ether double bond.

Oligomer and Film Characterization.   Brookfield viscosity measurements were taken on a Model RVTD digital readout viscometer. Samples for Instron testing were prepared on glass plates using 25 or 75 μm (1.0 or 3.0 mil) Byrd film applicator. Coatings for cure speed and MEK double rub (MEKDR) studies were prepared on aluminum Q-Panels® using a #40 wire wound rod (100 μm or 4.0 mil).

Coated substrates were cured by exposure to UV light from a Fusion Systems model F450 curing unit with a 120 w/cm (300 w/in) "D" lamp. This unit was mounted on a variable speed conveyor (4 to 75 m/min; 13 to 225 ft/min) and was capable of delivering a dose of 0.12 to 2.0 J/cm^2 in a single pass in air or nitrogen as measured with a UV Process Supply Compact Radiometer.

Tensile measurements were recorded using an Instron model 4201. Data was analyzed using Instron System 9 software. Tensile specimens were prepared by cutting 1.25 cm (0.5 in) wide strips of the coatings (75 μm; 3.0 mil) on glass plates. A 5.08 cm (2.0 in) gauge length was used with a crosshead speed of 2.54 cm (1") elongation per minute. The secant modulus at 2.5% elongation was recorded. A minimum of five tensile measurements were made for each sample.

Results and Discussion

Variation of Polyester Molecular Weight. A series of unsaturated polyester (UPE) oligomers were prepared by reacting maleic anhydride with 1,5-pentane diol. (Scheme

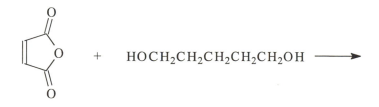

$$HO(CH_2CH_2CH_2CH_2CH_2O_2CCH = CHCO_2)_nCH_2CH_2CH_2CH_2CH_2OH$$

Scheme 2

2) The unsaturated polyesters were prepared by standard methods using butyl stannoic acid as the catalyst and removing the water of condensation with a xylene azeotrope. After reaching the desired acid value the xylene was removed under vacuum. The compositions prepared and their properties are summarized Table I. Decreasing the molar ratio of 1,5-pentane diol to maleic anhydride resulted in increased molecular weight of the unsaturated polyester. The theoretical molecular weight was calculated from Equation 2-85 in Odian's textbook with the assumption that the polymerization reaction had gone to completion (extent of reaction, p= 1).[16] This calculation provides an estimate of the maximum molecular weight for each oligomer. The actual molecular weight is probably somewhat less since all the oligomers had non-zero acid values. Brookfield viscosity an indirect measure of molecular weight increased linearly with the calculated molecular weight values.

Table I. Composition and Properties of Linear Unsaturated Polyesters with Different Molecular Weights

Reactant:	UPE A	UPE B	UPE C	UPE D
Maleic Anhydride	43.0%	44.7%	46.6%	47.6%
1,5-Pentane Diol	57.0%	55.3%	53.4%	52.4%
Properties:				
Mole Ratio (Diol:MA)	1.25	1.17	1.08	1.04
Viscosity (mPa·s)	3,340	6,100	23,400	40,080
Acid Value (mg KOH/g)	7.5	9.4	9.8	21.7
Mole Wt. (calc.)[16]	1773	2521	4957	10,378
Double Bond Eq. Wt. (calc.)	210	201	192	188

The unsaturated polyesters shown in Table I were formulated with triethylene glycol divinyl ether (DVE-3; ISP corp.) which has an equivalent weight of 101 at a 1:1 ratio of maleate double bond to vinyl ether double bond and 4% of 2-hydroxy-2-methyl propiophenone (Darocur 1173; Ciba-Giegy) was used as the photoinitiator. The formulations had the following Brookfield viscosities: A = 330 mPa·s; B: = 275 mPa·s; C = 655 mPa·s; and D = 960 mPa·s. Films were prepared from these coating formulations on aluminum Q-Panels with a #40 wire wound rod and cured at 0.5 J/cm². All the films exhibited > 200 MEK double rubs (MEKDR) and had a pencil hardness of "F" with a brittle film adhesion type failure. Films (3 mil) were also prepared and cured at 1 J/cm² on glass. The films could be removed from the glass substrate but were to brittle to obtain physical property measurement by testing on the Instron. All attempts to mount these films in the jaws of the machine resulted in destruction of the film and no tensile, elongation or modulus data could be obtained. Thus, very low equivalent weight unsaturated polyesters coupled with a low equivalent weight vinyl ether produce very brittle films.

Variation of Equivalent Weight. In an attempt to produce more flexible coatings, the UPE A formulation was modified to increase the equivalent weight by substituting succinic anhydride for a portion of the maleic anhydride. These unsaturated polyesters were prepared by the same method as above and the compositions and their properties are summarized in Table II.

Table II. Composition and Properties of Linear Unsaturated Polyesters with Different Equivalent Weights

Reactant:	UPE A	UPE E	UPE F
Maleic Anhydride	43.0%	28.6%	14.3%
Succinic Anhydride		14.6%	29.1%
1,5-Pentane Diol	57.0%	56.8%	56.6%
Properties:			
Viscosity (mPa·s)	3,340	3,050	2,800
Acid Value (mg KOH/g)	7.5	3.7	3.3
Double Bond Eq. Wt. (calc.)	210	316	634

The equivalent weight increased from 210 for UPE A to 316 for UPE E with one third of the maleic anhydride was replaced with succinic anhydride and 634 for UPE F with two thirds of the maleic anhydride is replaced with succinic anhydride. This change will effectively decrease the crosslink density of the final cured film. The Brookfield viscosities of the two new polyesters (E and F) are less than those from the first series. Since the molecular weights should be about the same (1760), it is not surprising that the Brookfield viscosities were relatively constant for all three unsaturated polyesters. The slight decreasing trend in viscosity with increasing succinic anhydride in the unsaturated polyester may be due to the increased free volume obtained when succinic anhydride is substituted for maleic anhydride.

The UV curable coating formulations were again prepared from these unsaturated polyesters, DVE-3 and 4% Darocur 1173 and are shown in Table III. Note that although UPE F has the lowest viscosity of the unsaturated polyesters, since it requires considerably less of the low viscosity (< 5 mPa•s) triethylene glycol divinyl ether to obtain 1:1 stoichiometry, Coating F has a higher viscosity. The reduced crosslink density of Coating F is shown by the lower value (131) for MEKDR compared to Coatings A and E (>200). Some increased flexibility has indeed been introduced into the network of Coating F as shown by the higher percent elongation of 7% but this increase is at the expense of a large decrease in modulus. For this linear unsaturated polyester simply reducing the equivalent weight does not produce the desired increase in flexibility and reduces both modulus and cure speed.

Table III. Coating Compositions and Their Properties

Component:	Coating A	Coating E	Coating F
UPE A	67.5		
UPE E		75.8	
UPE F			86.3
DVE-3	32.5	24.2	13.7
Darocur 1173	4.0	4.0	4.0
Viscosity (mPa·s)	330	340	750
Coating Properties: (1 mil film on aluminum Q-panel®)			
MEK Double Rubs (1 J/cm^2)	>200	>200	131
Physical Properties: (3 mil film on glass)			
Cure Dose (J/cm^2)	1.0	0.5	0.5
Tensile Strength (MPa)	(a)	4.1	1.5
Elongation (%)	(a)	3.0	7.0
Modulus (MPa)	(a)	153	19

(a) Too brittle to measure.

Variation of Maleate to Vinyl Ether Ratio. Variation of the stoichiometric ratio of vinyl ether double bond to maleate double bond should produce more flexible films due to plasticization by dangling end groups but also may reduce cure speed and mechanical properties. The curing of unsaturated polyester F with DVE-3 using 4% Darocur 1173

Table IV. Effect of Equivalent Ratio

Equivalent Ratio	0.5:1	1:1	1.5:1
Weight Percent DVE-3	7.4	13.7	19.3
Viscosity (mPa·s)	1195	750	470
MEKDR @			
1 J/cm^2	60	131	98
2 J/cm^2	65	132	80
3 J/cm^2	108	135	125

was studied at 1, 2 and 3 J/cm^2. The results are shown in Table IV. Unsaturated polyester F was chosen for this study because it had given a measurable value for MEKDR in the previous test. Because of its high equivalent weight (634), even when the vinyl ether to maleate equivalent raio was as high as 1.5:1, less DVE-3 was required on a weight percent basis than in any of the formulations A through E. The curing results indicate that doses higher than 3 J/cm^2 and a 1:1 ratio of vinyl ether to maleate are required to obtain a MEKDR value greater than 135 could be obtained. However, dosages of greater than 3 J/cm^2 are not of interest on a commercial basis. It also appears that 135 MEKDR may be close to the maximum value. The peeled coatings were flexible to the touch but had little or no tensile strength.

Photoinitiator Studies. A study with the UPE A/DVE-3 formulation using 2-hydroxy-2-methyl propiophenone as the photoinitiator showed that there was no difference in MEKDR values when the level of photoinitiator was varied from 0.5 to 3.0% even when dosages as low as 0.3 J/cm^2 were used. Thus, a photoinitiator level of 3.0% was chosen to evaluate some common photoinitiators using the UPE A system. The results are shown in Table V.

Table V. Screening Study of Common Photoinitiators

Photoinitiator	Cure Dose (J/cm^2)	MEK Double Rubs	Surface
Benzildimethyl ketal	1	100	Smudge
1-Hydroxycyclohexyl phenyl ketone	1	100	No damage
2-Hydroxy-2-methyl propiophenone	0.5	100	No damage
Isopropyl thioxanthone/Ethyl diethanol amine	1	80	All Removed
2,2-Diethoxy acetophenone	1	100	Smudge
Benzophenone/Ethyl diethanol amine	3	100	No damage
Camphoquinone/Ethyl diethanol amine	3	15	All Removed
Diphenyl (2,4,6-trimethyl benzoyl) phosphine oxide	0.5	100	No damage

Photoinitiators 2-hydroxy-2-methyl propiophenone and diphenyl-(2,4,6-trimethylbenzoyl) phosphine oxide both gave at least 100 MEKDR with no damage to the film surface when cured at a dosage as low as 0.5 J/cm^2. This preference in photoinitiators has been observed in other studies of the maleate/vinyl ether system.[10,11]

Investigation of Low Molecular Weight Diluents. An investigation of low molecular weight monomeric compounds as diluents was carried out by blending either diethyl

MONOMERIC DILUENTS

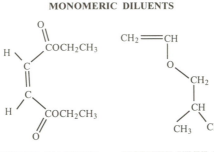

DIETHYL MALEATE   ISOBUTYL VINYL ETHER

maleate or isobutyl vinyl ether with UPE A and DVE-3. The objective was to increase the distance between crosslinks and thus improve the overall flexibility of the films. A three component Simplex type design was used to evaluate the utility of the diluents. The compositions were chosen so that the variation of the vinyl ether double bond to maleate double bond ratio was outside of the usual range of 0.5 to 1.5 which had been studied above. Darocur 1173 was used as the photoinitiator and the concentration was held constant at 3%. The cure speed was measured as the dosage necessary to obtain > 200 MEK double rubs and the pencil hardness was measured at this cure dosage.

The results for the formulations using diethyl maleate are summarized in Figure 1. The compositions of this design are well centered in the triangular graph. The viscosity of the formulations is decreased with increasing amounts of either DVE-3 or diethyl maleate. It can be seen that as the amount of diethyl maleate is increased the cure dosage necessary to obtain > 200 MEK double rubs also increases. This is probably due to the necessity to have high functionality to obtain cure speed. However, it is interesting to note that reasonable cure speed and pencil hardness can be achieved over a wide range of the vinyl ether double bond to maleate double bond ratio.

The results for the formulations containing isobutyl vinyl ether are shown in Figure 2. Due to the vinyl ether double bond to maleate double bond ratio constraint, the compositions used for this study are located in one corner of the triangular graph. It was unfortunate that compositions containing little or no DVE-3 were incompatible and could not be evaluated. Viscosities were not measured on these formulations. Cure dose increases as the amount of monofunctional diluent increases. Pencil hardness increased as the vinyl ether double bond to maleate double bond ratio increased. These results indicate that substitution of a monofunctional maleate or vinyl ether diluent results in reduced cure speed and pencil hardness.

Synthesis and Investigation of Urethane Vinyl Ether Oligomers. Another method of introducing flexibility into the films would be to use an oligomeric vinyl ether as one of the components. A difunctional vinyl ether oligomer was synthesized by reacting about 1 mol of a hydroxyalkyl vinyl ether with about 1.5 mol of a diisocyanate and about 1 mol of an oligomeric diol. This synthesis is shown in Scheme 3. The viscosity of this oligomer was about 3.7 million mPa·s corresponding to a molecular weight of about 2000 or an equivalent weight of about 1000. The blend of this oligomer with UPE A and 4%

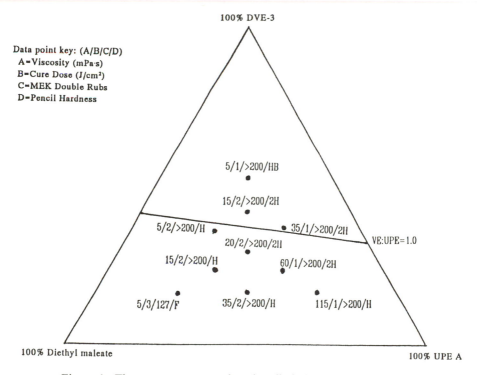

Figure 1. Three component study using diethyl maleate as a diluent.

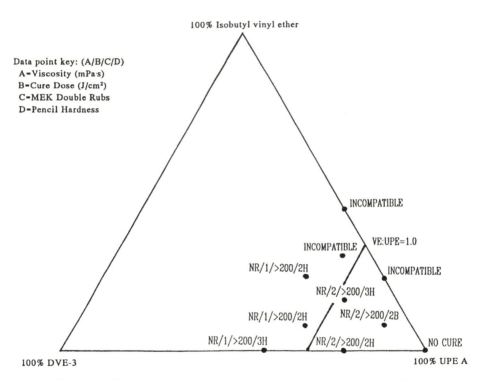

Figure 2. Three component study using isobutyl vinyl ether as a diluent.

$$2 \, CH_2\!\!=\!\!CHO\!-\!R\!-\!OH + 3 \, OCN\!-\!R'\!-\!NCO + 2 \, HO\!-\!R''\!-\!OH \longrightarrow$$

$$CH_2\!\!=\!\!CHO\!-\!R\!-\!O_2CNH\!\!\left[\!-\!R'\!-\!NHCO_2\!-\!R''\!-\!O_2CNH\!\right]_2\!\!R'\!-\!NHCO_2\!-\!R\!-\!OCH\!\!=\!\!CH_2$$

Scheme 3

Darocur 1173 had a viscosity of 260,000 mPa·s and cured at 1 J/cm^2 to give 175 MEK double rubs. A 3 mil film cured at 2 J/cm^2 had a tensile strength of 5.6 MPa, an elongation of 39% and a modulus of 27 MPa. This formulation was further diluted with a 1:1 stoichiometric mixture of DVE-3 and diethyl fumarate at a ratio of 90% formula to 10% diluent and cured at 2 J/cm^2. This produced little change in the mechanical properties. The tensile strength was 6.0 MPa, the elongation was 42% and the modulus was 25 MPa. A stoichiometrically balanced formulation of this oligomer (52.5%), butanediol divinyl ether (9.2%) and UPE A (38.3%) using 4% Darocur 1173 as the photoinitiator had a viscosity of 24,700 mPa·s and cured to > 200 MEK double rubs at 0.3 J/cm^2. A 3 mil film cured at 1 J/cm^2 had a tensile strength of 17 MPa, an elongation of 15% and a modulus of 140 MPa. The use of a urethane vinyl ether oligomer was found to be the best method of introducing flexibility into the cured vinyl ether/maleate films.

Conclusions

Maleate/vinyl ether formulations based on a model unsaturated polyester prepared from maleic anhydride and 1,5-pentane diol and triethylene glycol divinyl ether were studied. At molecular weights of less than about 10,000 the cured films were extremely brittle. When the equivalent weight of the unsaturated polyester was increased by replacing some of the maleic anhydride with succinic anhydride, measurable values for film elongation could be obtained but the cure speed was definitely slower. When either diethyl maleate or isobutyl vinyl ether were added as monofunctional diluents the cure dose needed to obtain 200 MEKDR was increased and the flexibility measured by pencil hardness increased as the amount of diluent was increased. A urethane vinyl ether was synthesized and used to replace DVE-3 and films with increased elongation were obtained at equivalent at dosages as low as 1 J/cm^2.

Acknowledgments

The author would like to acknowledge the technical assistance of E. Moschovis and the helpful discussions with S. Lapin. The author would also like to thank DSM Desotech, Inc. for permission to publish this work.

Literature Cited

1. Eaton, B.; Hanrahan, B. P.; Braddock, J. K., *RadTech '90 Conference Proceedings*, **1990**, *Vol* I, pp 384.
2. Klemarczyk, P.; Levandoski, S.; Okamoto, Y., *RadTech '92 Conference Proceedings*, **1992**, *Vol* I, pp 539.
3. Brautigam, R. J.; Lapin, S. C.; Snyder, J., *RadTech '90 Conference Proceedings*, **1990**, *Vol* I, pp 99.
4. Lapin, S. C.; Snyder, J. R., *RadTech '90 Conference Proceedings*, **1990**, *Vol* I, pp 410.
5. Woods, J. G.; Jacobine, A. F., *RadTech '92 Conference Proceedings*, **1992**, *Vol* I, pp 173.
6. Salamon, P. A.; Jacobine, A. F.; Glaser, D. M., *RadTech '92 Conference Proceedings*, **1992**, *Vol* I, pp 239.
7. Rakas, M. A.; Jacobine, A. F., *RadTech '92 Conference Proceedings*, **1992**, *Vol.* I, pp 462.
8. Noren, G. K.; Murphy, E. J., in *Radiation Curing of Polymeric Materials*, ACS Symposium Series No. 417, C. E. Hoyle and J. F. Kinstle, Ed.; American Chemical Society, Washington, D. C., 1990, pp 151.
9. Noren, G. K., *RadTech '90 Conference Proceedings*, **1990**, *Vol* I, pp 191.
10. Noren, G. K.; Tortorello, A. J.; Vandeberg, J. T., *RadTech '90 Conference Proceedings*, **1990**, *Vol* II, pp 201.
11. Schouten, J. J.; Noren, G. K.; Lapin, S. C., *RadTech '92 Conference Proceedings*, **1992**, *Vol* I, pp 167.
12. Shirota, Y., in *Encyclopedia of Polymer Science and Engineering*, 2nd ed., J. I. Kroschwits, J. I.; et al., Ed.; John Wiley and Sons, New York 1985, *Vol* 3, pp 327.
13. Johnson, S.; Schaeffer, W.; Hoyle, C. E.; Owens, J.; Shimose, M.; Sundell, P., *RadTech '94 Conference Proceedings*, **1994**, *Vol* 1, pp 194.
14. Li, T.; Luo, B.; Chu, G.; Hall, Jr., H. K., *J. Polym. Sci.; Part A: Polym Chem.*, **1990**, *Vol* 28, pp 1735.
15. Lapin, S. C., *Polymer Materials and Engineering Science*, **1989**, *Vol* 60, pp 233.
16. Odian, G. G., *Principles of Polymerization*, 3rd ed, John Wiley and Sons, New York, 1991, pp 79.

Chapter 11

Photoinitiator Free Polymerization of Maleimides and Vinyl Ethers

Charles E. Hoyle[1], Sonny Jönsson[2], Makoto Shimose[3], Jim Owens[1], and Per-Erik Sundell[4]

[1]Department of Polymer Science, University of Southern Mississippi, Box 10076, Hattiesburg, MS 39406–0076
[2]Fusion UV Curing Systems, AETEK UV Systems, 7600 Standish Place, Rockville, MD 20855–2798
[3]Nippon Steel Chemical Company, Ltd., East Wing, No. 2 Nippon Steel Building, No. 31–1, Shinkawa 2 Chome, Chuo-ku, Tokyo 104, Japan
[4]SSAB Tunnplåt AB, Borlänge, Sweden

The photopolymerization of mixtures of maleimides and vinyl ethers is shown to be an efficient, rapid process in the absence of external photoinitiators. Polymerization proceeds both in the presence and absence of oxygen. Films produced by the photopolymerization of maleimide/vinyl ether systems exhibit little absorbance at wavelengths greater than 300 nm. The thermal stability of these films are also excellent.

The use of photoinititators as integral components in UV curable systems is vital in providing a rapid and efficient method for producing radical or cationic species capable of initiating polymerization.[1] Both homolytic cleavage and hydrogen abstraction processes have been effectively used to generate radicals capable of initiating free-radical polymerization. Other processes have been used to produce protons or carbon centered cations for initiation of cationic polymerization. Of course, in both cases, it is necessary to have monomeric species capable of sustaining rapid chain growth processes in order to produce films at a fast rate. Although many monomer species have been incorporated into free-radical and cationic polymerizations, acrylates (free-radical) and vinyl ether or epoxy systems (cationic) have received the greatest attention in both basic studies and practical applications.[2] Recently,[3-8] there have been numerous references to the use of maleate (or fumarate) ester/vinyl ether comonomers in the development of UV curable systems which polymerize rapidly and yield crosslinked films (when multifunctional components are used) upon exposure to UV light. Of course, these systems are also constrained by the requirement of adding an external photoinitiator to absorb light and generate free-radical initiators. The copolymerization of the maleate ester and vinyl ether may be influenced by a charge-transfer (CT) interaction with the maleate ester as acceptor and vinyl ether as donor. Apparently, if any donor/acceptor CT complex is formed, it is not capable of efficiently absorbing light and initiating free-radical polymerization since exposure of maleate ester/vinyl ether mixtures to light in the absence of an added free-radical photoinitiator results in little polymerization. In addition to the commercially successful maleate (fumarate)/vinyl ether systems which use photoinitiators, numerous reports exist describing photoinitiated polymerization involving monofunctional monomer mixtures in the absence of added photoinitiators.[9-15]

In this paper, we report efforts to find donor/acceptor systems, comprised of at least one multifunctional monomer, capable of sustaining rapid free-radical polymerization without the need for external photoinitiators. Although we will include in this report comonomer systems which form ground state CT complexes, we stress that the primary mechanism for generating free-radical in each case may not be via excitation of ground state CT complexes.

Experimental

All maleimides were synthesized according to standard procedures.[16] Maleic anhydride, dimethyl maleate, and diethyl fumarate were purchased from Aldrich Chemical Co. and used as received. 1,4-Cyclohexanedimethanol divinyl ether (CHVE) and tetraethylene glycol divinyl ether (CHVE) were used as received from International Speciality Products. Bis(4-vinyloxybutyl) isophthalate (IPDBVE) and bis(vinyloxybutyl)succinate (SEGDVE) were obtained from Allied-Signal and used without further purification. All acrylates were used as received from either Aldrich Chemical Co. or Scientific Polymer Products. 2,2-Dimethoxy-2-phenylacetophenone (DMAP) was used as received from Ciba Specialty Chemicals.

A value for the polymerization enthalpy of 21.5 kcal/mole can be used to estimate percent conversion and rates for N-substituted maleimide/vinyl ether and maleic anhydride/vinyl ether copolymerizations. A value of 18.6 kcal/mole can be used for the enthalpy of polymerization of acrylate monomers to convert heat evolution data to percent conversion. Since the molar heats of polymerization for N-substituted vinyl ether copolymerization and acrylates vary by less than 20 percent, the exotherm data in the text are compared directly.

A modified Perkin Elmer DSC 2B was used to record exotherms for the monomer mixtures. The monomer mixtures (3 μL) were placed in crimped aluminum DSC pans using a calibrated microsyringe and degassed in the DSC for three minutes before irradiation. Polymerizations were carried out at 40 °C. Light from a 450 W medium pressure mercury lamp (Canrad Hanovia) was filtered through Pyrex before reaching the sample and reference cells. The light intensity (typically 13-20 mW cm^{-2}) was determined by placing a black DSC pan in the sample cell. The TGA unit employed was from TA Systems and the UV spectrometer from Perkin Elmer (Lambda 6). A Fusion 600D curing system was used to cure samples not exposed in the photo-DSC unit.

Results And Discussion

In considering the potential candidates for donor/acceptor photoinitiated polymerization in the absence of an added photoinitiator, we have evaluated a number of different types of donor and acceptor molecules (Table I). In this paper, we will focus on results for vinyl ether donors with the acceptors shown in Table I. However, before presenting any mechanistic results for particular systems, a brief overall evaluation of several combinations of monomer types is given in Table II for UV curing of thin monomer films on glass substrates with a Fusion Systems 600D UV curing unit. In all cases, the acceptors were monofunctional and donors difunctional. Precise intensity data for the UV curing process is not given since only relative conclusions are drawn from this "practical" evaluation. As the results in Table II show, maleimide/vinyl ether systems exhibit excellent curing properties. With this analysis in hand, attention is given to more precise rate data in order to substantiate our rather crude practical analysis of photoinitiator free systems. Acronyms and structures of all acceptor and donor monomers used in this study are shown in Diagram 1: Where appropriate, chemical names for the monomers used in this report are provided in the text or the experimental section.

N-Alkylmaleimide acceptor/ difunctional vinyl ether donor: It is certainly well known[17] that vinyl ethers are not capable of sustaining free-radical homopolymerization. However, there are numerous examples of free-radical copolymerization[18-20] of various monomer types with vinyl ethers. Maleimides,[21-25] on the other hand, are capable of free-radical homopolymerization. Although the homopolymerization rates of N-substituted maleimides are slow relative to acrylates, they have a strong tendency to participate in relatively rapid copolymerization with electron donor compounds such as vinyl ether or styryloxy monomers. In order to assess the relative ability of N-alkylmaleimides to participate in copolymerization with difunctional vinyl ether comonomers, exotherms of the photoinitiated polymerization of an equal mole percent N-hexylmaleimide (HM)/cyclohexyldimethanol divinyl ether (CHVE) mixture were compared with those of equal molar mixtures of dimethyl maleate (DMM)/CHVE and diethyl fumarate (DEF)/CHVE. Since the heats of polymerization of maleimide/vinyl ether systems are approximately equal to those of DMM/vinyl ether or DEF/vinyl ether systems, the results in Table III for mixtures exposed to equivalent (7.83 mW/cm^2) light intensities at 313 K both in the absence and presence of added photoinitiator can be easily compared. The maximum polymerization exotherm rates (H_{max}) for the HM/CHVE system with 1% photoinitiator were about 5 times greater than for the DMM/CHVE and DEF/CHVE mixtures. Perhaps most surprising are the results for the HM/CHVE system in the absence of any added photoinitiator. Not only did the maximum polymerization rate obtained for the photoinitiator-free HM/CHVE mixture far exceed the rates for photoinitiator- free DMM/CHVE and DEF/CHVE systems (under the conditions employed, exotherms were too low to measure for these systems), but the maximum rate of the photoinitiator free HM/CHVE mixture also exceeded that obtained for the DMM/CHVE and DEF/CHVE systems with 1 weight percent photoinitiator! The exotherm results for the HM/CHVE system in Table III substantiate the conclusion from Table II that maleimide/divinyl ether mixtures give viable cured films when exposed to typical UV curing light intensities.

In considering the origin of the results in Table III and the extraordinary rate exhibited by the HM/CHVE system, it is tempting to consider the role that ground-state charge transfer interaction between the donor and acceptor monomers in the

Table I	
Donors	**Acceptors**
Vinyl Ethers	Maleic Anhydride
p-Alkoxystyrenes	N-Arylmaleimides
N-Vinylpyrrolidone	N-Alkylmaleimides
N-Vinylformamides	Dialkyl Maleates
N-Vinylalkylamides	Dialkyl Fumarates

Table II			
Summary of Practical UV Curing Results: **Correlation with UV Spectral Results** • **Donors -- Difunctional** • **Acceptors -- Monofunctional** • **Air Not Excluded**			
Donor	**Acceptor**	**CT Complex (UV)**	**UV Curability***
Styryloxy	Maleic Anhydride	Very Strong	Low
Vinyl Ether	Maleic Anhydride	Strong	Good
Styryloxy	Maleimide	Moderate	Good
Vinyl Ether	Maleimide	Weak	Excellent
Vinyl Ether	Maleate or Fumarate	Very Weak	Poor-Good
*Curing conducted with Fusion 600D System *No photoinitiator added			

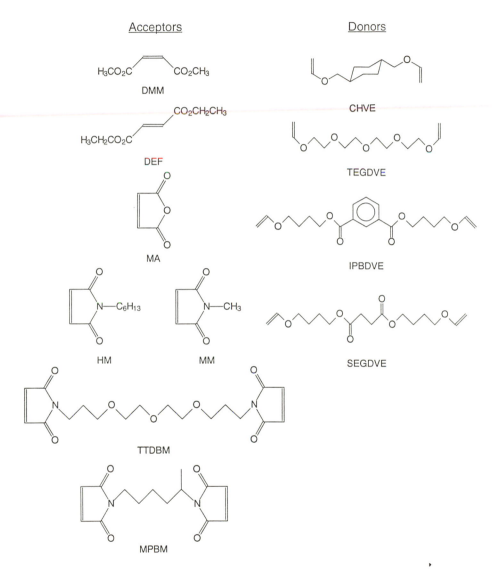

Diagram 1

initiation process may play. Indeed, UV spectroscopy of concentrated solutions of CHVE and HM indicate the presence of what is probably a relatively weak CT complex. Alternatively, the DEF/CHVE and DMM/CHVE mixtures show little tendency to form CT complexes, i.e., at best very, very weak complexes are formed. Based on these considerations, it would be of interest to obtain rate data for polymerization of divinyl ethers and a much stronger electron acceptor monomer, for example maleic anhydride: Such a system forms a stronger CT complex. Results are shown in Table IV for photoinitiated polymerization of equal molar mixtures of tetraethyleneglycol divinyl ether (TEGDVE) with N-hexylmaleimide (HM) and maleic anhydride (MA). It is quite obvious that the HM/TEGDVE system has a much higher degree of conversion and polymerizes at a much faster rate than the MA/TEGDVE system. Although the MA/TEGDVE system indeed exhibits the formation of what appears by UV spectroscopy to be a relatively strong absorbing CT complex, the rate of polymerization is relatively slow and the overall conversion low compared to the MA/TEGDVE mixture. One explanation for the remarkable ability of HM to initiate and participate in photo-polymerization with vinyl ethers might possibly be due to formation of an intermediate strength (not very weak as for the DMM/CHVE or DEF/CHVE systems or strong as for the MA/TEGDVE system) charge-transfer complex.[9-11] However, we should reiterate that we only observed the formation of a very weak CT complex between TEGDVE and HM, or CHVE and HM.

There is another explanation that we feel may explain the results obtained for the photoinitiator free polymerization of N-alkylmaleimide/vinyl ether systems. Consider the possibility that upon absorption of light, excited state N-alkylmaleimide abstracts a hydrogen from one of the components in the medium to yield a radical (s) capable of initiating the alternating copolymerization of the N-alkylmaleimide and vinyl ether functionalities. Accordingly, we tested the ability of N-alkylmaleimides to participate in excited state hydrogen abstraction reactions by obtaining polymerization results for N-alkylmaleimides as photoinitiators of acrylate polymerization. In order to best demonstrate the ability of N-alkylmaleimides to act as hydrogen abstraction type photoinitiators of acrylate polymerization, it is necessary to compare results between acrylates which are capable of rapid polymerization and which either have, or do not have, readily abstractable hydrogen atoms. 1,6-Hexanediol diacrylate (HDDA) is an excellent difunctional acrylate which polymerizes very rapidly via a free-radical process and has no readily abstractable hydrogens. Triethylene glycol diacrylate (TEGDA) and tripropylene glycol diacrylate (TPGDA) are examples of difunctional acrylates which have readily abstractable hydrogens. N-Methylmaleimide (MM) is chosen as the N-alkylmaleimide, since it is not capable of participating in an internal hydrogen abstraction process. Figure 1 shows exotherm curves for the photoinitiated polymerization of HDDA, TEGDA and TPGDA in the presence of 10 percent N-methylmaleimide, and no added conventional photoinitiator. As might be expected, no appreciable exotherm is recorded for HDDA. However, the presence of abstractable hydrogens in the diacrylate backbone of TEGDA and TPGDA give rise to efficient and fast polymerization. When molecules such as aliphatic amines, polyethyleneglycol, or multifunctional thiols are added to MM/HDDA solutions, rapid rates of polymerization are attained upon exposure to light (Figure 2). Similar results are obtained when these same hydrogen donating molecules are added to acrylate systems containing benzophenone, a known photoinitiator which requires a hydrogen donating coinitiator to function. The results in Figures 1 and 2 strongly suggest that

Table III		
Reactivity of Excited State Complex Effect of Electron Acceptor		
Acceptor	H_{max} (cal g^{-1}s^{-1})	
		1% DMAP
N-Hexyl Maleimide (HM)	0.56	1.68
Dimethyl Maleate (DMM)	0.00	0.34
Diethyl Fumarate (DEF)	0.00	0.31
I_o = 7.83 mW/cm^2, Filter: Pyrex, Temp.: 313K, N$_2$ Donor: CHVE		

Table IV				
Mixture	T_{max}(s)	H_{max} (cal g^{-1}s^{-1})	Area (cal g^{-1})	% Conversion
HM/TEGDVE	89	0.61	126	83
MA/TEGDVE	65	0.24	54.7	25
Pyrex - 13.6 mW/cm^2 T - 40°C				

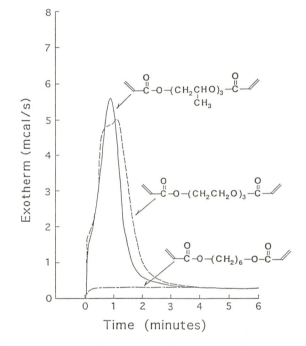

Figure 1. Exotherm curves for exposure of mixtures of 10 weight percent MM and diacrylate monomers (200 ppm hydroquinone inhibitor) to medium pressure mercury lamp through Pyrex filter ($I = 19.4$ mW cm^{-2}).

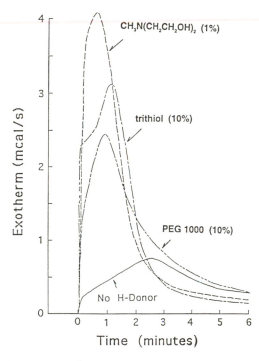

Figure 2. Exotherm curves for exposure of mixture of 10 weight percent MM and 1,6-hexanedioldiacrylate with concentrations (by weight) of added cosynergist to medium pressure mercury lamp through Pyrex (I = 14.1 mW cm^{-2}).

N-alkylmaleimides are capable of initiating free-radical polymerization when a coinitiator (species with readily abstractable hydrogen atoms) is present. Some preliminary laser flash photolysis results suggest that the hydrogen abstraction process of N-alkylmaleimides occurs from the triplet state formed upon excitation of N-alkylmaleimides. [In other work, we have found that N-arylmaleimide/difunctional vinyl ether mixtures are not photoreactive compared to N-alkylmaleimide/difunctional vinyl ether systems]. At this time, although we do not know for sure, it is tempting to speculate that hydrogen abstraction by excited triplet HM is responsible for generating the initiating radicals in N-alkyl maleimide/divinyl ether systems.

Finally, we should indicate that we have not ruled out the possibility that there is a contribution to initiating photopolymerization in maleimide/vinyl ether systems from an exciplex type complex between an excited state maleimide and ground state vinyl ether. A biradical formed from such a complex might initiate free radical polymerization in lieu of cyclization to form a 2 + 2 adduct. However, we note that at present we have no evidence for such a reactive exciplex.

Before presenting results on attempts to maximize the reactivity of maleimide/vinyl ether systems by altering the chemical structure of the two components, it is first necessary to compare the reactivity of maleimide/vinyl ether systems to the reactivity of a conventional acrylate system. If clear comparisons are to be made between vinylether/maleimide mixtures and conventional acrylates, results should be obtained both in nitrogen saturated and air saturated environments. Accordingly, Figure 3 shows exotherm curves of the photoinitiated polymerization of a HM/Di-(4-vinylglycol)butyl isophtalate (IPDBDVE) mixture in both nitrogen and air atmospheres. [The initial increase in the exotherm at very short times is probably a result of heat generated from exposure to the lamp and not polymerization.] The rather modest decrease in the polymerization exotherm rate maximum in air compared to nitrogen in Figure 3 is in stark contrast to the precipitous drop in the polymerization rate (Figure 4) of a comparable acrylate system in air (compared to nitrogen) of an equal molar HDDA/HA (hexylacrylate) mixture (Figure 4) with 1 weight percent of a common photoinitiator, 2,2-dimethoxy-2-phenylacetophenone (DMAP). Even after irradiation for several hundred seconds, the HDDA/HA/DMAP system shows little conversion, while the HM/IPDBDVE mixture produces a hard, crosslinked film under the same exposure conditions at relatively short exposure times. The origin of this oxygen effect may be due to a variety of factors including a decreased sensitivity to oxygen inhibitor in donor/acceptor systems, oxygen diffusion rates, etc.

Difunctional vinyl ether/difunctional N-maleimide. Up until this point, our results have centered on the reactivity of monofunctional maleimide divinyl ether mixtures. From Kloosterboer's[26] work for acrylate polymerization, it is known that the rate of polymerization of a free-radical process is increased dramatically as the functionality of the acrylate is increased. In order to enhance the polymerization rates of maleimide divinyl ether systems, it was decided to synthesize difunctional maleimides for copolymerization with difunctional vinyl ethers. The results in Table V indicate that the photoinitiated TTDBM [bismaleimide made from maleic anhydride and 4,7,10-

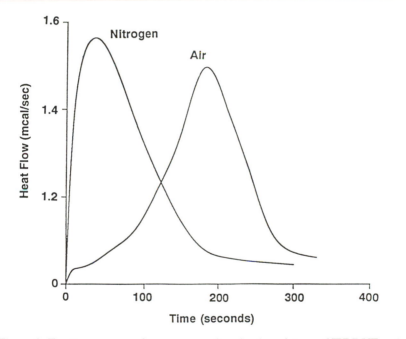

Figure 3. Exotherm curves for exposure of equimolar mixture of IPDDVE and HM in nitrogen and air to medium pressure mercury lamp through Pyrex (I≈20 mW cm⁻²).

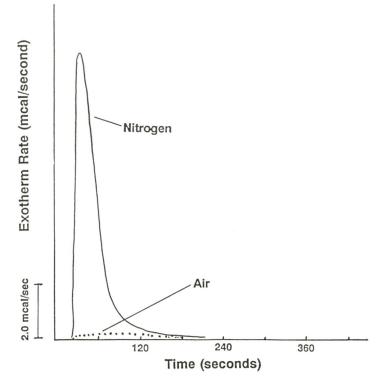

Figure 4. Exotherm curves for exposure of equimolar mixture of HA and HDDA in nitrogen and air with 1 weight percent DMAP photoinitiator to medium pressure mercury lamp through Pyrex ($I \approx 20$ mW cm^{-2}).

Table V		
Photopolymerization of Maleimides and Vinyl Ethers		
Maleimide + Vinyl Ether $\xrightarrow[\text{(PI)}]{h\nu}$ Polymer		
Monomers	H^1_{max} (cal g^{-1}s^{-1})	H^2_{max} (cal g^{-1}s^{-1})
HM/CHVE (5/5)	0.75	2.36
TTDBM/CHVE (5/5)	1.49	2.48
TTDBM/TEGDVE (5/5)	1.43	2.96
HDDA/HA (5/5)	0.00	2.61
Light Intensity: 13.5 mW/cm^2, Sample Amount: 3 μL, Filter: Pyrex, Temperature: 40 °C [1]No photoinitiator; N_2 [2]1 weight percent DMAP added; N_2		

trioxa-1,13-tridecanediamine] polymerization of equimolar mixtures of the bismaleimide and two divinyl ethers in the absence of an added photoinitiator are faster (judged by the exotherm peak maximum) than for the HM/CHVE system. The increase for the TTDBM/CHVE versus the HM/CHVE system in the absence of a photoinitiator could be due to hydrogen abstraction of the α-methyl hydrogen on the diether backbone of the TTDBM, as well as the difunctionality of TTDBM. Also included in Table V are results for the polymerization of the equimolar mixtures of HM/CHVE, TTDBM/CHVE, and TTDBM/TEGDVE with an added conventional photoinitiator (DMAP). The results are quite interesting since the polymerization rates are not significantly faster when DMAP is included in the mixture. Additionally, there is little difference in the peak maximum of the sample with the monofunctional maleimide (HM/CHVE) and the samples with the difunctional maleimide [TTDBM/CHVE and TTDBM/TEGDVE}.

Figure 5 shows UV absorption spectral results for exposure of a mixture of TTDBM and a divinyl ether made from succinic acid and 1-butanol vinyl ether (SEGDVE) to an unfiltered medium pressure mercury lamp for various time periods. A dramatic decrease in the maleimide group absorption (peak maximum near 300 nm) occurs as polymerization proceeds, indicating that the maleimide functional group is rapidly consumed during polymerization. Other results[27] indicate that the vinyl ether functionality is also consumed rapidly by the polymerization process, concomitant with the loss of maleimide functionality. If the system used to obtain the results in Figure 5 is coated on a quartz substrate and exposed to a medium pressure mercury lamp (Fusion system D bulb 45m/min; Intensity = 1.4W/cm^2) most of the maleimide absorbance is lost (Figure 6) in 1 pass (total of 20 msec exposure). From additional investigations conducted in our lab, the vinyl ether groups also experience high conversions in excess of 80 percent. The matrices obtained in 1 pass are highly crosslinked non-tacky films. This is significant since it clearly demonstrates that photoinitiator free bismaleimide/divinyl ether resins cure at very rapid rates: The rates are comparable to those obtained for acrylate resins with photoinitiators using the same lamp system.

In view of the surprising results obtained for the monofunctional maleimide HM/IPDBDVE system in Figure 3 for polymerization in air, an equimolar mixture of CHVE and MPBM [bismaleimide made from maleic anhydride and 2-methyl-1,5-pentanediamine] was polymerized via a mercury lamp source in the absence and presence of air (Figure 7). As might be expected from the earlier results in Figure 3, the difunctional maleimide/divinyl ether system exhibits a remarkably high polymerization rate in the presence of oxygen (air).

In order for maleimide/vinyl ether photoinitiator free photopolymerization to be useful, it is important that the cured films have good thermal/UV stability. Since there are no small molecule photoinitiators added to the uncured mixture initially, there is no residual small molecule photoinitiator present in the final crosslinked film. This accounts for the enhanced UV stability we have observed for cured maleimide/vinyl ether films. In addition, TGA thermograms of photocured films of the MPBM/CHVE mixture (Figure 8) exhibit excellent thermal stability, with decomposition occurring at higher temperatures than for a simple UV cured HDDA film with 3 weight percent DMAP photoinitiator. (Such thermal stability would be

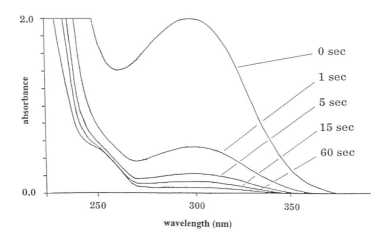

Figure 5. UV absorption spectra of a thin film of an equimolar mixture of TTDBM/SEGDVE as a function of exposure time to a Pyrex-filtered medium pressure mercury lamp (I ≈ 19 mW cm^{-2}).

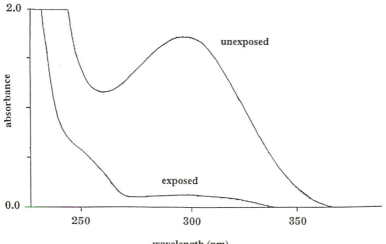

Figure 6. UV absorption spectra before and after ~20 msec exposure of a thin film of an equimolar TTDBM/SEGDVE mixture to a Fusion D bulb (I = 1.4 W cm^{-2}; line speed = 45 m min^{-1}; total exposure ≈ 28 mJ cm^{-2}).

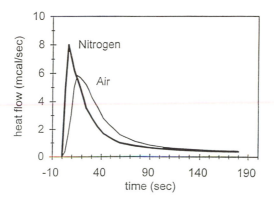

Figure 7. Exotherms of equimolar mixtures of MPBM and CHVE in air and nitrogen exposed to a medium pressure mercury lamp through a Pyrex filter (I ≈ 19 mW cm⁻²).

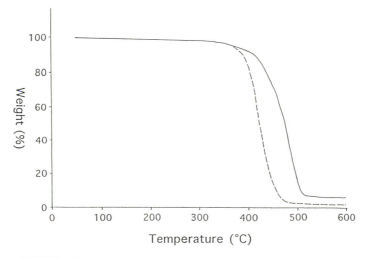

Figure 8. TGA of cured films (cured by exposure to Fusion 600 D model lamp) of an equimolar MPBM/CHVE mixture (—) and a HDDA film with 3 weight percent DMAP photoinitiator (---). Light intensity approximately same as for Figure 6 with multiple exposures for each system.

expected from earlier results for thermal curing of similar systems.[28]) Other properties of UV cured maleimide/vinyl ether films are under current investigation in our lab.

Conclusions

In this paper, it has been demonstrated that photocurable systems can be made from mixtures of maleimides and vinyl ethers without the need for addition of external small molecule photoinitiators. Superior rates relative to acrylate polymerization are attained in an air environment. The ability of excited state maleimides to generate radicals capable of initiating polymerization via hydrogen abstraction has been shown. The rapid disappearance of absorbance of the maleimide band upon exposure of maleimide/vinyl ether systems to light demonstrates the loss of maleimide groups during polymerization. Moreover, this loss of absorbance at wavelengths greater than 300 nm is instrumental in generating films with decreased sensitivity to long term exposure to irradiation. This has implications for outdoor weatherability of UV cured films.

Acknowledgments

The support of Fusion UV Curing Systems and Nippon Steel Chemical Co. is gratefully acknowledged.

Literature Cited

1.   Fouassier, J.P.; "Photoinitiation, Photopolymerization, and Photocuring," **1995**, Hanser Publishers, New York, NY.

2.   Fouassier, J.P.; Rabek, J.; eds. *Radiation Curing in Polymer Science and Technology, I-IV,* , **1993**, Elsevier Applied Science, New York, NY.

3.   Zahora, E.P.; Lapin, S.C.; Noren, G.K.; Wehman, E. *Rad Tech '94 Conf. Proc.*, **1994**, 621.

4.   Lapin, S.C.; Noren, G.K.; Schouten, J.J.; *Rad Tech Asia '93 Conf. Proc.,* **1993**, 149.

5.   Noren, G.K.; Tortorello, A.J.; Vandeburg, J.T. *Rad Tech '90 North America Conf. Proc. II,* **1990**, 201.

6.   Schouten, J.J.; Noren, G.K.; Lapin, S.C. *Rad Tech '92 North America Conf. Proc., I,* **1992**, 167.

7.   Noren, G.K.; Schouten, J.J.; Lapin, S.C. *Proc. XX Water-Borne, Higher-Solids and Powder Coatings Symposium,* **1993**, 256.

8.   Lapin, S.C.; Noren, G.K.; Julian, J.M. *PMSE,* **1995**, 72, 589.

9.   Li, T.; Lee, C.; Hall, H.K. *Adv. Polym. Sci.,* **1990**, 97, 1-339 and references therein.

10.   Lee ,C.; Hall, H.K. *Macromolecules,* **1989**, 22, 21.

11.   Li, T.; Luo, B.; Chu, G.; Hall, H.K. *J. Polym. Sci., Polym. Chem.,* **1990**, 28, 1735.

12.   Barton, J.; Capek, I.; Arnold, M.; Rätzsch, M. *Makromol. Chem.,* **1980**, 181, 241.

13. Sadhir, R.K.; Smith, J.D.B. *J. Polym. Sci. Polym. Chem.,* **1992**, 30, 589.
14. Wipfelder ,E.; Heusinger, E. *J. Polym. Sci. Chem. Ed.,* **1978,** 16, 1779.
15. Li, T.; Pan, J.; Zhang, Z. *Makromol. Chem.,* **1989,** 190, 1319.
16. Wang, Z. Y. *Synth. Commun.,* 1990, 20, 1607.
17. Olson, K.G.; Butler, G.B. *Macromolecules,* **1984,** 17, 2486.
18. Hwo, J.C.H.; Miller, L. *J. Polym. Sci.,* **1961,** 55, 197.
19. Mayo, F.R.; Lewis, F.M.; Walling, C. *J. Amer. Chem. Soc.,* **1948,** 70, 1529.
20. Noma, K.; Niwa, M. *Kobunshi Kagaka,* **1972,** 29, 52.
21. Aida, H.; Takase, I.; Nozi, T. *Makromol. Chem.,* **1989,** 190, 2821.
22. Lin ,Y.; Huang, S.J. *Polym. Prep.,* **1989,** 248.
23. Zott, H.; Heusinger, H. *Eur. Polym. J.,* **1978,** 14, 89.
24. Wipfelder , E.; Heusinger, *J. Polym. Sci., Polym Chem. Ed.,* **1978,** 16, 1779.
25. Yamada, M.; Kakase, I.; Koutou, N. *Polym. Lett.,* **1968,** 6, 883.
26. Boots, H. M. J.; Kloosterboer ,J. G.; van der Hei, G. M. *Brit. Polym. J.,* **1985,** 17, 219.
27. Decker, C.; Morel, F.; Jönsson, S. ;Clark, ;S.C.; Hoyle, C.E. *PMSE,* **1996,** 95, 198.
28. Smith, M.A.; Do, Ch. H.; Wagener, K.B. *Polym. Prep.,* **1988,** 29(1), 337.

Chapter 12

UV Cross-Linking of Thiol-ene Polymers:
A Rheological Study

Bor-Sen Chiou, Robert J. English, and Saad A. Khan[1]

Department of Chemical Engineering, North Carolina State University, Raleigh, NC 27695–7905

An *in situ* dynamic rheological technique was used to characterize the UV cross-linking behavior of several thiol-ene systems. The cross-linking rate increased with increasing UV light intensity; however, the fractal dimensions at the gel point were comparable for different UV light intensities, indicating that the structures at the gel point were independent of UV light intensity. In addition, the application of pulse or continuous modes of irradiation had no effect on the evolving rheological properties during UV curing. In fact, the gel times and fractal dimensions at the gel point, determined from the rheological data, were comparable regardless of which mode of irradiation was used. The extent of reaction of the thiol-ene polymerization was determined by using a real-time Fourier transform infrared (FTIR) technique. This was combined with the dynamic rheological data to correlate changes in the elastic modulus with extent of reaction.

Traditionally, UV curable polymers have been utilized as coatings for wood and vinyl floors, but their applications have increased dramatically over the last twenty years to encompass many diverse areas, including optical fiber coatings (*1*), adhesives (*2*), disc replications (*3-5*), and microelectronics (*6*). This widespread use of UV cross-linked systems is attributed to their rapid, energy efficient curing and their solvent free, one piece formulations. Typically, UV curable systems require only a small fraction of the power normally utilized in thermally cured systems and their solvent free nature offers an environmentally safer alternative.

Commonly used monomers for UV curing include acrylates (*7*), styrene/unsaturated polyesters (*8,9*), and thiol-ene compositions (*10-12*). Currently, acrylate-functional systems constitute a major share of the UV curable polymers market, mainly due to their rapid curing via free radical chain polymerization.

[1]Corresponding author

150

Another factor contributing to their preeminence in the UV curing industry is the wide range of material properties which can be obtained by varying both the acrylate functionality and the chemical architecture between the acrylate groups. However, there has been a resurgence in interest for nonacrylate photopolymers, due to the increasingly stringent environmental and toxicological regulatory issues associated with acrylates (*13*). This has led to renewed interest in thiol-ene compositions, which offer a toxicologically safer alternative to acrylate-functional systems.

The flexibility of formulating thiol-ene photopolymers has prompted their use in novel applications, ranging from conformal coatings for printed circuit boards to primary coatings for optical fibers. Thiol-ene silicones and formulations based on allyl terminated urethane elastomers have low glass transition temperatures and low moduli, both desirable characteristics for reducing microbending losses when used as primary coatings for optical fibers (*1*). Thiol-ene photopolymers have also been used as adhesives, especially for bonding optics, because shrinkage and residual stress after curing is low. One major advantage of thiol-enes over acrylate-functional systems is that they are not inhibited by oxygen (*13*). In acrylate-functional systems, oxygen may deactivate a propagating chain, thus reducing the rate of polymerization and requiring longer UV exposure times.

Tailoring suitable compositions of UV curable systems for specific properties are usually carried out on an empirical basis. This is due to the lack of understanding between how the molecular architecture of the monomers and the curing conditions affect the final properties of the cured polymer. To increase this understanding, studies have focused on the kinetic aspects of the photoinitiated polymerization of UV curing systems. Thiol-ene polymerizations have been studied by both differential scanning calorimetry (*10,11*) and Fourier transform infrared spectroscopy (FTIR) (*12*). The final mechanical properties of the cured polymers have also been investigated (*12,14*). However, there has been a lack of studies which focus specifically on the evolving rheological properties during UV cross-linking. The evolution of the rheological properties directly reflects the kinetics of cross-linking, as dictated by both the extent and density of cross-links. This ultimately determines the final material properties of the cured polymer. Processing parameters, such as UV radiation intensity and UV curing temperature, affect the rheological properties during UV curing, which are crucial for optimizing the curing process.

The evolving rheological properties during UV curing is usually studied using a pulse mode of UV irradiation because of the lack of an *in situ* technique. The sample is initially irradiated with a pulse of UV light and the rheological properties are subsequently measured. This process is continued until the rheological properties, during the transformation from a liquid to a solid sample, is fully mapped out. However, in most industrial applications, UV curing occurs by a continuous mode of irradiation, where the sample is continually irradiated. The rheological properties measured by using a pulse mode may be different than those measured by using a continuous mode because dark curing, the further cross-linking due to trapped radicals in the absence of UV radiation, may occur while the rheological properties are being measured.

Additional insights into the fundamental aspects of UV curing may be gained by correlating rheological properties with extent of reaction obtained from infrared spectroscopy. By combining the time-dependent elastic modulus data from rheology with the extent of reaction with time data from FTIR, the relationship between rheological properties and material chemistry can be determined. This can be utilized to screen out possible polymers for specific applications. On a more fundamental level, these correlations can provide a basis for determining the effects of chemistry, such as chain length and functionality, on the rheological properties during UV curing.

Rheology, by itself and in combination with other methods, therefore provides a powerful and unique approach to study the various fundamental and applied aspects of UV cross-linking. In this chapter, we attempt to illustrate this point using model and commercial systems. Specifically, we utilize an *in situ* technique (*15,16*) to monitor the effects of UV light intensity on the evolving rheological properties during the UV curing of a commercially available thiol-ene optical adhesive, NOA 61 (Norland Corporation, New Jersey). We compare how pulse and continuous modes of UV irradiation affect their rheological properties. We also correlate the rheological properties, obtained by utilizing the *in situ* technique, with the extent of reaction, measured by using real-time FTIR, for a trimethylolpropane tris(2-mercaptoacetate) (trifunctional thiol)/trimethylolpropane diallyl ether (difunctional allyl) model system.

Experimental

Rheological Tests. The *in situ* monitoring of the photo-cross-linking process was conducted using a Rheometrics mechanical spectrometer (RMS 800) in conjunction with specially designed parallel plate fixtures (Figure 1). Quartz glass plates, each 20 mm in diameter, were held onto both the top and bottom fixtures by removable screws. A section of the top fixture was cut away, allowing the incident UV radiation to be directed onto the sample, via a mirror inclined 45° to the upper plate. The quartz plates were transparent to UV radiation and allowed the sample to be cured in the rheometer.

UV radiation was generated by a 200 W Oriel mercury lamp. A narrow band interference filter maintained the wavelength of the incident UV radiation at 365 nm, while neutral density filters were used to attenuate the light intensity. The intensity was measured using an International Light 1400A radiometer prior to each experiment. Radial variation in the incident light intensity was adjusted to less than ~ 7% in all cases. We examined the effects of two different light intensities, 0.02 and 0.17 mW cm^{-2} on the UV curing of the NOA 61 (Norland) samples. The thickness for all NOA 61 samples was 0.6 mm and all samples were cured at room temperature (25°C).

Dynamic rheological tests were used to monitor the evolving cross-linking structure during UV curing. In dynamic tests, a sinusoidal strain, γ, deformation

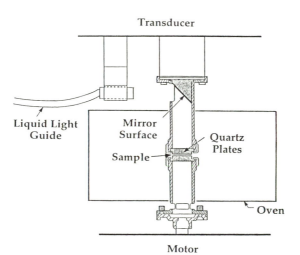

Figure 1. Schematic diagram of specially designed fixtures used to measure, *in situ*, the evolving rheological properties during UV curing. The sample is placed between the two quartz plates and the incident UV radiation exiting the liquid light guide is reflected by the mirror through the plates and onto the sample.

$$\gamma = \gamma_0 \sin \omega t \tag{1}$$

was applied to the sample at a frequency ω and a strain amplitude γ_0. The resulting stress, τ_{yx}, given by

$$\tau_{yx} = G' \gamma_0 \sin \omega t + G'' \gamma_0 \cos \omega t \tag{2}$$

can be decomposed into two parts, one in phase with the deformation, characterized by the elastic modulus, G', and one out of phase with the deformation, defined by the viscous modulus, G''. The elastic modulus measures the amount of energy stored by the sample during one cycle and is very sensitive to the microstructure of the sample. The viscous modulus measures the amount of energy dissipated during one cycle. Both the elastic and viscous moduli are extremely sensitive to the degree of cross-linking and provide a signature of the state of a material.

The pulse mode experiments were conducted by placing the sample between the quartz plates and exposing it to 30 second bursts of UV radiation. Dynamic rheological tests were then applied to the sample and this process was repeated until the sample had passed its gel point and became highly cross-linked.

For the continuous mode, we utilized a dynamic rheological technique, Fourier transform mechanical spectroscopy (FTMS) (17,18), which provided a powerful method for monitoring, simultaneously, the evolving dynamic moduli at several frequencies during the course of photo-cross-linking. In this technique, an oscillatory strain, γ, was applied to the sample, such that

$$\gamma = \sum_{i=1}^{m} \gamma_i \sin(\omega_i t) \qquad \omega_1 = \omega_f, \, \omega_2 = n_2 \omega_f, \, \dots \, \omega_8 = n_8 \omega_f \tag{3}$$

The resulting stress was measured, and a discrete Fourier transform was performed to obtain the elastic and viscous moduli. The experimental variables in FTMS are the fundamental frequency, ω_f, and the strain amplitudes, γ_i, at each frequency, ω_i. Each of the other frequencies are harmonics (integer multiples) of the fundamental frequency. The fundamental frequency was set at 1 rad/s, while the harmonics were chosen to be 2, 5, 10, 25, 40, 50, and 60 rad/s. The summation of the strain amplitudes at each frequency was below the linear viscoelastic limit of the NOA 61 sample.

Dynamic rheological measurements have recently been used to accurately determine the gel point (19). Winter and Chambon (20) have determined that at the gel point, where a macromolecule spans the entire sample size, the elastic modulus (G') and the viscous modulus (G'') both exhibit the same power law dependence with respect to the frequency of oscillation. These expressions for the dynamic moduli at the gel point are as follows:

$$G'(\omega) = S \ \Gamma(1\text{-}n) \ \cos(n\pi/2) \ \omega^n \qquad (4)$$

$$G''(\omega) = S \ \Gamma(1\text{-}n) \ \cos(n\pi/2) \ \tan(n\pi/2) \ \omega^n \qquad (5)$$

$$0 < n < 1 \qquad (6)$$

where ω is the frequency of oscillation, S is the gel stiffness, Γ is the gamma function, and n is the relaxation exponent. It follows from equations 4 and 5 that the loss tangent, tan δ, which is defined as G''/G', becomes independent of frequency at the gel point:

$$\tan \delta = \frac{G''}{G'} = \tan(n\pi/2) \qquad (7)$$

Therefore, the values of tan δ determined at different frequencies of oscillation, when plotted as a function of time, should converge at the gel time, t_{gel}.

Real-Time FTIR. For our IR studies, we utilized a stoichiometrically equivalent amount of a trifunctional thiol, trimethylolpropane tris(2-mercaptoacetate), with a difunctional allyl, trimethylolpropane diallyl ether. The thiols were protected from oxidative polymerization by the addition of hydroquinone. The monomers and hydroquinone were purchased from Aldrich Chemicals and were used as received. This formulation was mixed for five minutes and then a commercial photoinitiator, Esacure TZT (Sartomer Inc.), which contained a blend of methyl benzophenones, was added at a level of 1.0% by weight of monomers to the formulation. Stirring was maintained for a further five minutes following the addition of the photoinitiator. The final formulation contained 2.0% by weight of hydroquinone. The samples were prepared prior to each experiment in order to ensure reproducibility of sample history.

The apparatus for obtaining real-time IR spectra during UV curing consisted of a liquid light guide and two BaF_2 crystals. The UV radiation, generated by a 200 W Oriel Mercury lamp, was introduced into the FTIR sample chamber by the flexible liquid light guide. This light guide was held at a distance of 6 cm from the BaF_2 crystals to ensure total UV light coverage of the 25 mm x 4 mm BaF_2 crystals. It was also positioned at an angle of 15° so that it did not block the path of the IR beam. The thiol-ene sample was placed between the two BaF_2 crystals along with a 0.05 mm teflon spacer and then the whole unit was placed in a Spectra-Tech Presslock Holder. The wavelength of the UV light was maintained at 365 nm by a narrow band interference filter, while the intensity, which was measured by an International Light 1400A radiometer, was held constant at 0.15 mW cm^{-2}.

An FTIR spectrometer (Nicolet Magna-IR System 750) and a DTGS-KBr detector was used to obtain the IR spectra for the thiol-ene sample. Sixteen scans, with a resolution of 4 cm^{-1}, were averaged for each spectrum, with each spectrum totalling 24 seconds to complete. The spectra were measured one after another to ensure the sample was analyzed in real-time. Either the S-H stretching vibrations of the thiol functional groups at 2570 cm^{-1} or the C=C stretching vibrations of the allyl functional groups at 1646 cm^{-1} can be used to calculate the extents of reaction during UV curing. We chose the thiol absorption because it was stronger than the allyl absorption. The aromatic out-of-phase C-H deformation vibrations from the methyl benzophenone photoinitiators at 830 cm^{-1} was utilized as an internal standard since this peak remained unchanged throughout the whole process.

The extent of reaction of the thiol (2570 cm^{-1}) functional groups at any time during UV curing can be determined by first integrating the unreacted thiol peak at the start of the experiment. At time t, the area of the peak can be integrated and the extent of reaction at that time can be determined as follows:

$$x(t) = \frac{A(0) - A(t)}{A(0)} \tag{8}$$

where $x(t)$ is the extent of reaction at time t, $A(0)$ is the normalized area (with respect to 830 cm^{-1}) of the initial thiol peak, and $A(t)$ is the normalized area of the thiol peak at time t.

Since we wanted to correlate the elastic modulus with extent of reaction, the experimental conditions for the dynamic rheological measurements had to be as close as possible to those used when obtaining the IR spectra. Therefore, we used FTMS to measure the dynamic moduli during UV curing of samples 0.05 mm thick and also used an UV light intensity of 0.15 mW cm^{-2}. In order to check whether the UV radiation penetrated the entire sample thickness, we varied the sample thickness from 0.05 mm to 0.10 mm. We found no changes in the gel time, indicating that our sample thickness of 0.05 mm was within the penetration depth of the UV radiation. The thiol-ene samples used for both IR and rheological studies were cured at 25°C.

Results and Discussion

Rheological tests are able to monitor the buildup of the cross-linking structure during UV curing since the elastic modulus (G′) is very sensitive to changes in microstructure. Using our *in situ* technique, we measured the evolving rheological properties during the UV curing of an NOA 61 sample. Figure 2 shows the behavior of the elastic (G′) and viscous modulus (G″) as a function of the frequency of oscillation at three different UV exposure times. At t = 1280 s, before the sample has reached its gel point, the viscous modulus is larger than the elastic modulus by an order of magnitude and the moduli are not parallel, indicating that the sample is still a liquid. After the sample is exposed to 1660 s of UV radiation, both the elastic and viscous moduli scale with frequency to the same power, satisfying the Winter criterion for the gel point (equations 4 and 5). This means that at this time, the sample has reached its gel point. For a time much greater than the gel time, t = 2780 s, the elastic modulus becomes independent of frequency and larger than the viscous modulus throughout the frequency range, both of which are characteristics of a highly cross-linked network. Note that the viscous modulus increases by about two orders of magnitude while the elastic modulus increases by up to four orders of magnitude (at the lower frequencies) during UV curing, thereby revealing the sensitivity of G' to the evolving structure.

We examine the effects of varying the UV light intensity on the kinetics of cross-linking in Figure 3, where we plotted the elastic modulus for two intensities as a function of the UV exposure time. The sample irradiated with the higher UV intensity exhibits a larger slope and therefore, a faster cross-linking rate. This is due to the larger number of photoinitiators excited by the higher UV light intensity, which subsequently initiates more polymerization reactions. In addition, as the UV intensity is increased, the gel time decreases because of the higher cross-linking rates at the higher intensities.

We are also interested in studying the effects of using pulse versus continuous modes of irradiation on the evolving rheological properties during UV curing. Bair et al. (*2*) showed that when intermittent doses of UV radiation were applied to an adhesive sample, the glass transition temperature and final conversion were lower than those obtained from the same dose using continuous irradiation. Therefore, dark curing may have played a significant role in the kinetics, and ultimately, the final material properties of the UV cured polymer. In Figure 3, we compare the elastic modulus using the pulse and continuous modes of irradiation for the two different intensities. The modulus, using the pulse mode (30 second pulses), matches quite well with the modulus determined by using the continuous mode from the beginning of the reaction to a time slightly past the gel point. There appears to be no dark curing during this period for the NOA 61 sample. This agrees with a study by Khan (*21*), which showed that when an NOA 61 sample was cured to a state before the gel point was reached, both the elastic and viscous moduli at different frequencies of oscillation remained constant over time, even after 21 hours. An electron spin resonance study done on the sample also indicated no free radicals were present, which precludes dark curing since this mechanism relies solely on the presence of free radicals. Although pulse or continuous modes of irradiation did not affect the rheological properties for this particular sample, this may not be the case for other samples where dark curing occurs.

Gel Point.   A significant aspect of cross-linking polymerizations is the determination of the gel point, the point at which an infinitely large macromolecule appears. This gel point can be accurately measured by performing dynamic rheological tests and applying the Winter criterion. Using equations 4 and 5, we can determine the gel time by examining, on a log-log plot, when the elastic and viscous moduli as a function of the frequency of oscillation become parallel. Such a plot is shown in Figure 4, where the elastic and viscous moduli at the gel point from both the pulse and continuous modes of irradiation are plotted as a function of the

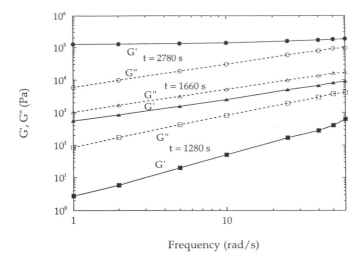

Figure 2. The elastic (G′) and viscous modulus (G″) as a function of the frequency of oscillation at three different UV exposure times. At t = 1660 s, G′ and G″ are parallel, indicating the sample has reached its gel point. Note that G′ is initially smaller than G″ throughout the frequency range (t = 1280 s), but becomes larger than G″ and independent of frequency towards the end of the reaction (t = 2780 s), indicating that the sample has been transformed from a liquid to a highly cross-linked network.

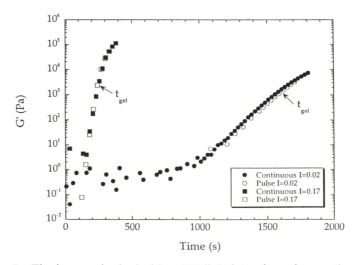

Figure 3. The increase in the incident UV light intensity enhances the cross-linking rate, as indicated by the larger slopes of the elastic modulus (G′) plotted as a function of UV exposure time. The use of either the pulse or continuous modes of irradiation do not affect the rheological properties for the NOA 61 sample. Here, the frequency of oscillation is 10 rad/s and the intensities have units of mW cm^{-2}.

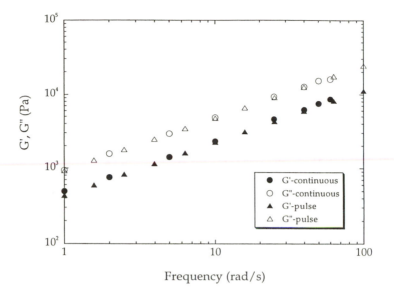

Figure 4. The elastic (G′) and viscous modulus (G″) at the gel point match well for samples irradiated by using the pulse or continuous modes.

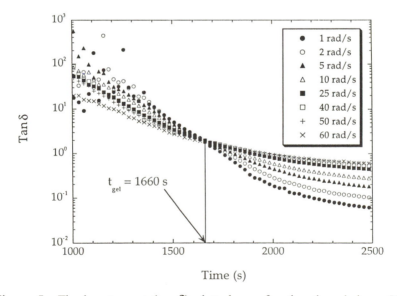

Figure 5. The loss tangent (tan δ) plotted as a function time during a UV cross-linking reaction. All the values of tan δ for the different frequencies intersect at the gel time of 1660 s, indicating that the sample has reached its gel point. Here, the incident UV light intensity is 0.02 mW cm^{-2}.

frequencies of oscillation. The moduli match for both modes of irradiation (continuous and pulse), consistent with the conclusion that the different modes do not affect the rheological properties. However, the particular time when the dynamic moduli are parallel during the course of the experiment is often difficult to discern precisely. Therefore, we can determine the gel point more easily by using equation 7, which states that the loss tangents (tan δ's) for all frequencies converge at the gel time. Such a plot of tan δ as a function of the UV exposure time is shown in Figure 5, where we find the values of tan δ to intersect at a time of 1660 s, corresponding to the gel time, t_{gel}. Using this technique, the gel times were determined for both the pulse and continuous modes at the two different intensities. For an intensity of 0.02 mW cm^{-2}, the gel times for the pulse and continuous modes were 1650 $\pm$50 s and 1620 $\pm$30 s, respectively. The corresponding gel times for the higher intensity of 0.17 mW cm^{-2} were calculated to be 250 $\pm$20 s for the continuous mode and 250 $\pm$20 s for the pulse mode. We observe no difference in gelation times for the two modes within experimental error.

We also determined the relaxation exponents, n, at the gel point from tan δ plots and by using equation 7. For an intensity of 0.02 mW cm^{-2}, the values of n were found to be 0.70 $\pm$0.01 for the continuous mode and 0.71 $\pm$0.01 for the pulse mode, indicating that the properties of the critical gel do not depend on the mode of UV irradiation. Similar results were obtained for a higher intensity of 0.17 mW cm^{-2}, where n was determined to be 0.67 $\pm$0.01 for the continuous mode and 0.69 $\pm$0.02 for the pulse mode. Within experimental error, it seems that values of the relaxation exponents were comparable, regardless of whether pulse or continuous modes were used on the sample. Also, the UV light intensity had little effect on the values of the relaxation exponents determined for the NOA 61 samples.

The power-law variation of the dynamic moduli at the gel point has led to theories suggesting that the cross-linking clusters at the gel point are self-similar or fractal in nature (22). Percolation models have predicted that at the percolation threshold, where a cluster expands through the whole sample (i.e. gel point), this infinite cluster is self-similar (22). The cluster is characterized by a fractal dimension, d_f, which relates the molecular weight of the polymer to its spatial size R, such that

$$R^{d_f} \sim M \tag{9}$$

The fractal dimension measures how open or packed a structure is; lower fractal dimensions indicate a more open system, while higher fractal dimensions indicate a more packed system (22). Theories relating the fractal dimension to the relaxation exponent, n, have been put forward and these are based on whether the excluded volume of the polymer chains is screened or unscreened under conditions near the gel point (23). It is known that the excluded volume of a polymer chain is progressively screened as its concentration is increased, the size of the chain eventually approaching its unperturbed dimensions. Such screening is expected to occur near the gel point, since cross-linked clusters of varying sizes are present and the situation is similar to that encountered in a polymer melt, as far as the excluded volume effects are concerned (23). For the case when the excluded volume is fully screened, the relaxation exponent, n, can be related to the fractal dimension, d_f, at the gel point by (23),

$$n = \frac{d(d + 2 - 2d_f)}{2(d + 2 - d_f)} \tag{10}$$

where d is the space dimension, which in this case is 3. By applying Equation 10, fractal dimensions of 1.9 for both modes were calculated for the low UV intensity and fractal dimensions of 1.8 for both modes were calculated for the high UV intensity. The comparable fractal dimensions for both intensities suggest that the cross-linked structures produced at the gel point are similar and independent of both intensity and modes of irradiation.

Activation Energy. The gel times, determined by dynamic rheological tests, can also be utilized to calculate an apparent activation energy. We can obtain the gel times over the temperature range of interest and if the extent of reaction at these temperatures are constant, an apparent activation energy can be determined. First, the polymerization reaction can be represented by a generalized kinetic expression of the type: (24)

$$\frac{dx}{dt} = k\, f(x) \ ; \qquad k = A\, \exp(-E/RT) \tag{11}$$

where x is the extent of reaction, t is the time of reaction, k is the rate constant, f(x) is an arbitrary functional form for the extent of reaction, T is the temperature of the reaction, A is the preexponential factor, and E is the apparent activation energy. Equation 11 can be integrated from the onset of the cross-linking reaction (time t = 0, x = 0) to the gel point (t = t_{gel}, x = x_c) to obtain

$$\ln t_{gel} = \ln(\int_0^{x_c} dx/f(x)) - \ln(A) + E/RT \tag{12}$$

Thus, a semilogarithmic plot of the gel time as a function of 1/T should be linear, with the slope corresponding to the apparent activation energy. We have determined the gel times for a temperature range of 25°-50° C for a thiol-ene system consisting of stoichiometrically equivalent amounts of a trifunctional thiol, trimethylolpropane tris(2-mercaptoacetate), and a trifunctional allyl monomer, triallyl isocyanurate. In this system, we also added 0.31% by weight of hydroquinone, to prevent premature polymerization, and 1.0% by weight of a commercial photoinitiator, Esacure TZT.

The semilogarithmic plot of gel times as a function of 1/T for this system is illustrated in Figure 6 and an apparent activation energy of 6.6 kcal/mol is obtained.

Correlation of Rheology with FTIR. The IR spectra of the trimethylolpropane tris(2-mercaptoacetate)/ trimethylolpropane diallyl ether system at three UV exposure times is presented in Figure 7. At the beginning of the polymerization, the thiol (2570 cm^{-1}) and the allyl (1646 cm^{-1}) peaks are distinct. However, as the polymerization progresses, the peaks decrease in size, as more of the thiol and allyl functional groups are reacting to form cross-links. Finally, towards the end of the reaction, both the thiol and allyl peaks have diminished in size as most of the functional groups have reacted. The aromatic out-of-phase C-H deformation peak (830 cm^{-1}), which was used as the internal standard, did not change throughout the reaction.

The elastic modulus (G'), obtained using the *in situ* rheological technique, and the thiol extent of reaction, measured by real-time FTIR, as a function of the UV exposure time is shown in Figure 8. The elastic modulus does not reach an appreciable value until after about 800 s of UV exposure or about 70 % conversion of the thiol functional groups. This seems reasonable because the gel time occurs after 930 seconds of UV exposure and before the gel point is reached, there is no macromolecule which spans the sample size. Therefore, during this period, the sample contains smaller cross-linked clusters which may not be of sufficient size to give rise to a large elastic modulus. However, once the gel point has been reached, the elastic modulus increases very rapidly as the sample becomes more densely cross-linked. Similar results have been obtained for a PDMS system (25), where an appreciable elastic modulus was not measured until slightly before the gel time. Note that the conversion of the functional groups is occurring the fastest towards the beginning of the reaction, when the value of the elastic modulus is still very low. Therefore, the maximum rate of conversion and the maximum rate of increase in the elastic modulus do not necessarily match.

When the elastic modulus is plotted as a function of converson, as shown in Figure 9, one can see more clearly how the elastic modulus does not start increasing until higher levels of conversions are attained. However, after this point, the elastic modulus increases very rapidly. In fact, for less than twenty percent conversion of the functional groups, the elastic modulus increases by more than three orders of magnitude, showing the sensitivty of the elastic modulus for probing the microstructure of the cross-linking networks.

Conclusions

We have established that an *in situ* rheological technique provides a powerful way to monitor the UV curing of thiol-ene systems by obtaining such parameters as gel time, fractal dimension, and activation energy from the rheological properties. We found that the cross-linking rate increased with increasing UV light intensity, resulting in

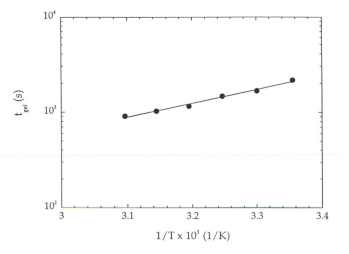

Figure 6. The semilogarithmic plot of gel times as a function of 1/T for the trimethylolpropane tris(2-mercaptoacetate)/triallyl isocyanurate system is used to determine the apparent activation energy. From the slope, the apparent activation energy was found to be 6.6 kcal/mol.

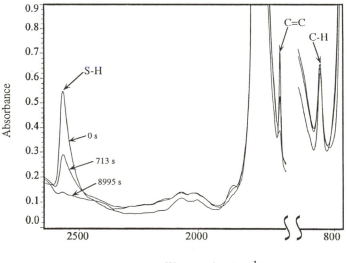

Figure 7. IR spectra of the trimethylolpropane tris(2-mercaptoacetate)/ trimethylolpropane diallyl ether system for three different UV exposure times (0 s, 713 s, 8995 s) showing S-H stretching (2570 cm^{-1}), C=C stretching (1646 cm^{-1}), and aromatic out-of-phase C-H deformation (830 cm^{-1}) peaks. The thiol peak is monitored to calculate the extent of reaction while the C-H deformation peak is used as an internal standard.

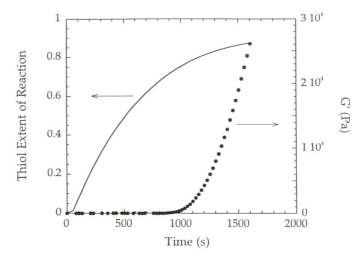

Figure 8. The extent of reaction as a function of UV exposure time, collected from real-time FTIR, along with the elastic modulus (G′) as a function of UV exposure time, measured by the *in situ* rheological technique, shows that an appreciable elastic modulus appears only after high extents of reactions are reached.

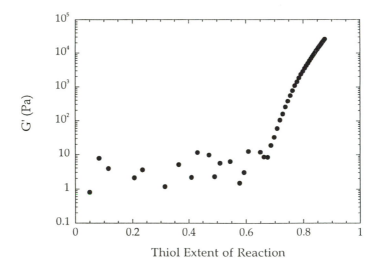

Figure 9. The plot of the elasic modulus (G′) as a function of the extent of reaction shows more clearly that once the elastic modulus is detected, it increases at a rapid rate. In fact, G′ increases by more than three orders of magnitude for less than 20% change in conversion of the thiol functional groups.

decreased gel times. The relaxation exponents, and consequently the fractal dimensions at the gel point, were found to be the same for both intensities studied, indicating that the cross-linked structures produced at the gel point were independent of UV light intensity.

We also compared the effects of using either a pulse or continuous mode of irradiation on the rheological properties of the UV cross-linking polymers. The two modes of irradiation had no effect on the evolving dynamic moduli since the moduli obtained from both modes matched quite well. The gel times and relaxation exponents obtained from using the different modes were also comparable, suggesting that the rheological properties were independent of the irradiation modes. The reason for this may be that no dark curing occurred for the NOA 61 sample.

We also correlated the elastic modulus (G′) with the extent of reaction by combining data obtained from rheology and FTIR. We found that the elastic modulus did not show an appreciable increase until a short time before the gel point of the sample had been reached. However, in the vicinity of the gel point and beyond, the elastic modulus increased significantly with conversion. Information relating modulus to extent of reaction, which plays a critical role in the application of these systems, can therefore be readily obtained using our approach.

Acknowledgments

We thank the American Chemical Society Petroleum Research Fund and a GAANN Fellowship for providing financial support for this work.

Literature Cited

1)   Blyler, L.L., Jr.; Dimarcello, F.V. In *Encyclopedia of Physical Science & Technology*; Meyers, R.A., Ed.; Academic Press: New York, NY, 1987, 11; 637-647.
2)   Bair, H.; Blyler, L.L., Jr. *Proc. 14th NATAS Conf.* **1985**, *5*, 392.
3)   Bowman, C.N.; Carver, A.L.; Kennett, S.L.; Williams, M.M.; Peppas, N.A. *Polym. Bull.* **1988**, *20*, 329.
4)   Bowman, C.N.; Carver, A.L.; Kennett, S.L.; Williams, M.M.; Peppas, N.A. *Polymer* **1990**, *31*, 135.
5)   Kloosterboer, J.G.; Lippits, G.J.M. *J. Imag. Sci.* **1986**, *30*, 177.
6)   Kloosterboer, J.G. In *Advances in Polymer Science*; Dusek, K., Ed.; Springer-Verlag: Berlin, 1988, 84; 1-61.
7)   Tryson, G.R.; Shultz, A.R. *J. Polym. Sci., Polym. Phys. Ed.* **1979**, *17*, 2059.
8)   Kinkelaar, M.; Hsu, C.P.; Lee, L.J. *SAMPE Quarterly* **1990**, *21*, 40.
9)   Hsu, C.P.; Lee, L.J. *Polymer* **1991**, *32*, 2263.
10) Hoyle, C.E.; Hensel, R.D.; Grubb, M.B. *Polym. Photochem.* **1984**, *4*, 69.
11) Hoyle, C.E.; Hensel, R.D.; Grubb, M.B. *J. Polym. Sci., Polym. Chem. Ed.* **1984**, *22*, 1865.

12) Jacobine, A.F.; Glaser, D.M.; Grabek, P.J.; Mancini, D.; Masterson, M.; Nakos, S.T.; Rakas, M.A.; Woods, J.G. *J. Appl. Polym. Sci.* **1992**, *45*, 471.

13) Jacobine, A.F. In *Radiation Curing in Polymer Science and Technology*; Fouassier, J.B., Rabek, J.F., Eds.; Elsevier Science Publishers Ltd.: New York, NY, 1993, 3; 219-268.

14) Rakas, M.A.; Jacobine, A.F. *J. Adhesion* **1992**, *36*, 247.

15) Khan, S.A.; Plitz, I.M.; Frantz, R.A. *Rheol. Acta* **1992**, *31*, 151.

16) Chiou, B.S.; English, R.J.; Khan, S.A. *Macromolecules* **1996**, *29*, 5368.

17) Holly, E.E.; Ventkataraman, S.K.; Chambon, F.; Winter, H.H. *J. Non-Newtonian Fluid Mech.* **1988**, *27*, 17.

18) In, M.; Prud'homme, R.K. *Rheol. Acta* **1993**, *32*, 556.

19) Winter, H.H. In *Encyclopedia of Polymer Science and Engineering*; Kohlmetz, E., Levy, C., Walter, P.; John Wiley & Sons: New York, NY, 1989, Supplement Volume; 343-351.

20) Winter H.H.; Chambon, F. *J. Rheol.* **1986**, *30*, 367.

21) Khan, S.A. *J. Rheol.* **1992**, *36*, 573.

22) Feder, J. *Fractals*; Plenum Press: New York, NY, 1988.

23) Muthukumar, M. *Macromolecules* **1989**, *22*, 4656.

24) Oyanguren, P.A.; Williams, R.J. *J. Appl. Polm. Sci.* **1993**, *47*, 1361.

25) Venkataraman, S.K.; Coyne, L.; Chambon, F.; Gottlieb, M.; Winter, H.H. *Polymer* **1989**, *30*, 2222.

Emerging Applications

Chapter 13

Electrochemical Deposition of Photoresists

K. Chandrasekaran[1], M. J. Hill[1], W. L. Hamilton[2], H. V. Nguyen[3],
and R. W. Collins[3]

[1]Research and Development Division, DuPont, Inc., Towanda, PA 18848–9784
[2]Research and Development Division, DuPont, Inc., Wilmington, DE 19898
[3]Materials Research Laboratory, The Pennsylvania State University,
University Park, PA 16802

Spectroscopic ellipsometry is a non-destructive, interface sensitive, *in situ* technique for interface characterization. Time resolved ellipsometric spectroscopy was used to determine the mechanism of electrochemical deposition of photoresists on copper electrodes under potentiostatic, anodic conditions. Nucleation of photoresist deposition occurs randomly. During the early stages of nucleation the semi-spherical particles are separated by about 100 A. The deposits tend to grow like "pillars" up to 50 A. Further growth of the "pillars" lead to coalescence of the photopolymer deposits. Surface roughness decreases with increase in film thickness. A smooth film is produced when the film thickness is above 175 A. Thin electrodeposited photopolymer films were found to be fairly conducting.

Photoresists used for printed wiring board manufacturing are generally negative working dry films coated on a polyester substrate, dried and covered with polyolefin. Photoresists consist of film forming polymers, photopolymerizable acrylic, allylic or epoxy type monomers, photoinitiators, photosensitizers, colorants, adhesion promoters, etc.(1). Polyolefin is removed from the dry film sandwich and the film is laminated onto a copper substrate along with the polyester in a hot roll laminator, and exposed with a phototool to cross link the material. The polyester is removed and the unexposed polymer is washed off in a developing solvent. Generally, the film forming material is an acid-containing acrylic polymer, and the photopolymer film is developed in an aqueous sodium carbonate solution.

Conformance of dry film to the copper surface is a concern, particularly with stiff films. One way to improve conformance is to coat a low viscosity photopolymer liquid directly on the substrate. Several coating techniques are available to deposit photoresists directly onto metal substrates. These techniques involve dissolving/dispersing the ingredients in a solvent, coating the solution on the substrate and removing the solvent. Some of the material is lost during the process and the application yield is low. Also, removal and disposal of the solvent from the coating is a concern. Electrochemical deposition of paints and polymers from aqueous solution is done directly on the substrate, and the application yield is high(2). The electrochemically deposited film is almost dry, and minimum solvent removal is required.

Mechanism of electrochemical deposition of photoresists can be determined by monitoring the kinetics of film growth on the substrate under electrochemical conditions. Most of the surface analysis techniques involve examining the surface under vacuum. The substrate is removed from the solution, dried and placed in a high vacuum chamber. A high energy radiation is used to analyze the surface. It is generally assumed that these changes do not affect the interface. Solid-electrolyte interface can be very different from the solid-vacuum interface. Therefore, the results of vacuum surface science studies cannot always be extended to represent the solid-liquid interface. An *in situ* analytical method is preferred.

FTIR reflection spectroscopy is an excellent technique for characterizing a metal-organic film interface. Electrochemical deposition is generally done in an aqueous medium. Water absorbs strongly in the infra red region. In order to minimize infra red light absorption by water, a thin cell is used. Electrolyte diffusion is limited under these conditions and the kinetics are altered. Ellipsometric spectroscopy using visible light offers an advantage over infra red reflection spectroscopy because light absorption by the electrolyte is minimal.

When polarized light reflects from a surface, changes occur in the amplitudes and phases of the orthogonal electric field components resolved into the p and s directions (parallel and perpendicular to the plane of incidence, respectively). These changes lead to a change in the polarization. Both the thickness and optical properties of any layers present at the interface affect the change. In reverse, the optical properties and the thickness of the materials present at the interface can be determined by monitoring the amplitude and phase changes that occur upon reflection(3,4).

One of the advantages of ellipsometric spectroscopy is its sensitivity. Sub-monolayer coverages of films can be studied owing to the ability to measure phase changes to within 0.01°, or better. In addition, ellipsometric spectroscopy can be used in a variety of adverse environments including air, plasmas and liquids. Time resolved ellipsometry can be used to determine the kinetics and mechanisms of deposition of thin films on surfaces subjected to high voltages such as might occur in some electrodeposition processes.

Time resolved ellipsometry is an ideal technique to investigate the mechanism of nucleation and film growth of electrochemically deposited films.

Experimental

The electrolyte, AN-100, is an aqueous solution of a polymerizable unsaturated resin partially neutralized with triethylamine. Synthesis of the polymer and the electrolyte are described below.

A mixture comprising of 40 parts by weight of methyl methacrylate, 40 parts by weight of butyl methacrylate, 20 parts by weight of acrylic acid and 2 parts by weight of azobisisobutylnitrile was added dropwide to 90 parts by weight of propylene glycol monomethyl ether at 110° C over a period of 3 hours. Solutions were purged with nitrogen before addition and a nitrogen atmosphere was maintained during addition. After an hour, a 10% solution of azobisdimethylvaleronitrile in propylene glycol monomethyl ether was added over a period of 1 hour. To the solution, 24 parts by weight of glycidyl methacrylate, 0.12 parts by weight of hydroquinone and 0.6 parts by weight of tetraethylammonium bromide were added(5). The polymer was separated and dried. 15 parts by weight of the polymer was dissolved in 75 parts by weight of deionized water and 10 parts by weight of propylene glycol monomethyl ether. Triethylamine was used to neutralize 30% of the acid. Hydroxyisobutylphenone and diethylthioxanthone were used as photoinitiators.

Particle size of the micelle was determined by light scattering. Distance between the copper electrodes was maintained at 1 cm. After electrochemical deposition the electrode was removed from the cell and dried at 100° C for 10 min to remove water. Electrochemical deposition was done initially under galvanostatic conditions with the applied current density of 30 - 60 A per square foot, and then changed to potentiostatic conditions after achieving a required film thickness.

The ellipsometer used in this study is described elsewhere(3). It consists of a Xenon light source, a monochromator, a polarizer, a sample holder, a rotating analyzer and a photomultiplier detector (Figure 1). An electrochemical cell with two windows is mounted at the center. The windows, being 120° apart, provide a 60° angle of incidence for the ellipsometer. A copper substrate and a platinum electrode function as anode and cathode respectively. Both are connected to a DC power supply. The system is automated with a personal computer to collect all experimental data during the deposition. Data analysis is carried out by a Fortran program run on a personal computer.

Anionic polymer(s) and photosensitizer(s) dissolved in a water - alcohol mixture were used as the medium. A standard solution was diluted by 2000 times to reduce light scattering by the polymer particles and to reduce the current density in the working range.

A smooth electrode surface for the time resolved ellipsometric spectroscopy was prepared as follows: a 7059 glass slide was degreased by soaking in methanol for one hour, dried and placed in a sputter deposition chamber. High purity(99.999%) copper was sputtered in an argon atmosphere of 5 mTorr. The thickness of the copper was determined to be in the range of 4000 - 4500 A.

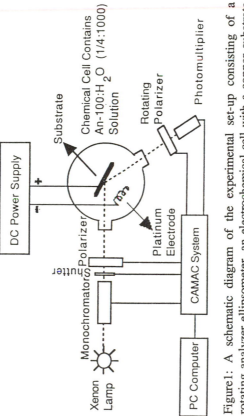
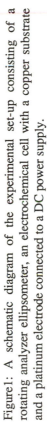

Figure1: A schematic diagram of the experimental set-up consisting of a rotating analyzer ellipsometer, an electrochemical cell with a copper substrate and a platinum electrode connected to a DC power supply.

Results and Discussion

Electrochemical Deposition

The polymer in the electrolyte forms micelle-like particles in the aqueous medium. Core of the micelles is primarily a hydrophobic component of a polymer and the shell is made of hydrophilic acrylic acid groups. Amine containing polymers are used for cathodic deposition. The micelle contains the polymeric binder, photochemically crosslinkable functional groups and photoinitiators. The particle size was determined to be in the range of 100 - 300 A using light scattering under the experimental conditions. The particles are negatively charged. The stability of the particles increases when the carboxylic acid is partially neutralized with a base. When acid is added to protonate the carboxylate groups the polymer particles tend to settle or deposit on available surfaces. The same thing can be accomplished by the addition of multivalent metal ions. For example, addition of zinc salts precipitates the resin.

Negatively charged micelles migrate to the positively charged electrode. The surface charge of the micelle is neutralized near the surface by the electrochemically generated hydrogen ion or metal ion in the case of anodic deposition or hydroxyl ions in the case of cathodic deposition. Solubility of neutralized micelles in the aqueous medium is less and the micelles tend to coalesce on the electrode surface. One of the advantages of electrochemical deposition is smooth film surface. If the electrode surface is rough, initial electrodeposition rate is different at different parts of the electrode due to diffusion in the local micro environment resulting in different rate of deposition and consequently the surface becomes smooth. Similarly pin holes are healed during the coating process. The deposited polymer film is a dielectric. The rate of deposition decreases as the deposited film thickness increases. Also, the catalytic nature of the film surface is different from that of the metal surface for the generation of hydrogen ions or metal ions.

This polymer contains carboxylic as well as acrylic functionality. The acrylic part of the polymer undergoes light induced polymerization. Use of this photoreactive polymer eliminated the need for separate acrylic type polymerizable monomer(s).

The charge on the outer surface of the micelle can be controlled by the amount of acrylic acid present in the main polymer and by partial neutralization. It is possible to prepare the polymer with amine containing monomers. Acidification of the polymer then forms particles with positive charge on the outside. Depending on the surface charge on the particles, the polymer can be anodically or cathodically deposited on the electrode surface. Positively charged polymer particles are cathodically deposited and negatively charged particles are anodically deposited. Anodic deposition is preferred from aqueous medium. Over-voltage required for anodic reactions in aqueous medium is larger than the over voltage required for the cathodic reactions. Equilibrium potential for hydrogen evolution in an aqueous acidic solution is 0.0 V. Even on non-catalytic surfaces hydrogen evolution tends to dominate in an aqueous medium during cathodic deposition.

Hydrogen bubble formation during nucleation results in pin holes in the film. Anodic deposition eliminates this problem. The equilibrium potential for oxygen evolution from water is 1.23 V. Over-voltage for oxygen evolution is even higher on non-catalytic surfaces. Other electrochemical reactions tend to dominate oxygen evolution. The oxidation potential for copper is 0.44 V. When copper is used as an anode, the electrode tends to dissolve before oxygen evolution. Copper ions released from the electrode complex with the carboxylic acid groups on the polymer particle surfaces, and the metallized particles deposit on the electrode.

Figure 2 shows the reaction for the anodic electrodeposition. The anodic micelle has carboxylic acid groups in the outer shell and is neutralized by the hydrogen ions or metal ions. Hydrogen ions are generated by the electrolytic oxidation of water at the copper electrode surface. Small amount of copper ions are also produced by the dissolution of the electrode under anodic conditions. When enough carboxylate groups are converted to uncharged carboxylic acid groups, the micelle collapses and coalesces on the electrode surface. As the coating thickness builds up water oxidation can occur on the copper electrode surface or on the polymer film surface. Since the water content is low on the electrode surface water oxidation is presumed to occur on the polymer film surface. Since the electrode is positively charged hydrogen ions are repelled and migrate to the film surface. Diffusion of water to the electrode surface and the diffusion of hydrogen ions from the electrode surface to the film surface control the micelle deposition rate. As the film thickness increases, diffusion decreases and the micelle deposition rate decreases. Water oxidation at the electrode surface is not by the diffusion of water and hydrogen ions. However, the polymer film is a poor conductor and the potential drop within the film increases with thickness limiting the electrode potential at the interface. Hence, water oxidation at the surface of the electrode also decreases with film thickness.

Figure 3 shows the cathodic reaction. The counter ion, protonated trialkyl ammonium ion, migrate through the ion selective membrane to the cathode. Hydroxyl ions are generated at the cathode where they reduce the amine ions to free amines.

The electrodeposited film thickness as a function of time is shown in figure 4 for various current densities. An induction period of up to 1 min is observed. This is attributed to change in pH near the electrode surface. Initially, a linear increase in film thickness is observed which tends to level off at longer times. Film thickness does not increase linearly with current density even at low film thickness. Diffusion of micelles to the electrode surface is slow compared to coalescence. At 40 mA/dm^2 a film growth rate of 125 microns per amphere per minute is observed which is the best yield indicating the diffusion and consumptin are similar under these conditions. Since the electrode is held under galvanostatic conditions the equilibrium potential is dependent on the set current. A lower equilibrium potential is expected for low current density and a high equilibrium potential is expected for high current densities. The difference in film thickness under equilibrium is attributed to the difference in the energy supplied.

In order to confirm the plating uniformity a copper panel electrode was prepared by etching depressions on the copper surface. These depressions are

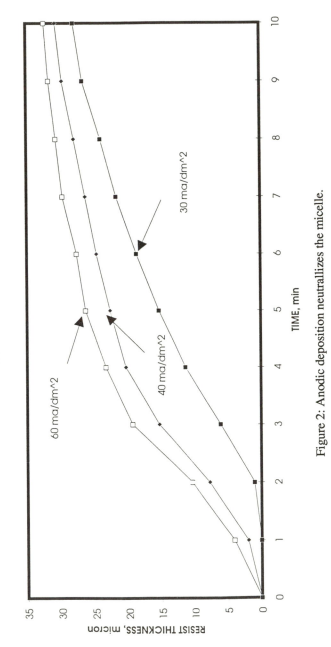

Figure 2: Anodic deposition neutrallizes the micelle.

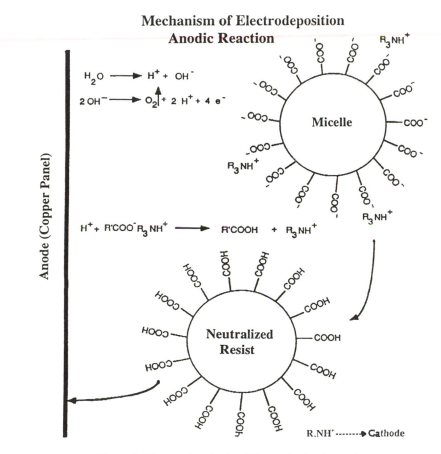

Figure 3: Electrochemical cell for cathodic deposition.

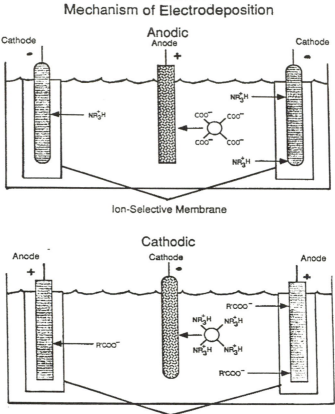

Figure 4: Electrochemical deposition rate as a function of time at various current densities.

about 17 microns deep. After anodic electrochemical deposition smooth film surface was observed. The panel was imagewise exposed and developed in sodium carbonate solution. Developed lines on the rough surface is shown on Figure 5. There is excellent conformation of the resist to the surface.

One of the advantages of electrodeposited photoresists is the absence of polyester film on the surface of the film during imaging exposure. The phototool is in close contact with the photopolymer film compared to the dry film where the phototool and the photopolymer film are separated by a polyester sheet. Better contact results in improved phototool reproduction. The lines/space reproduction of 125 micron at various exposures is given in Figure 6. Image size is almost constant for 300% change in exposure energy whereas the image size of the dry film varies by more than 10% under similar conditions. Consequently, finer lines can be resolved with electrodeposited photoresists(Figure 7).

Ellipsometic spectroscopy

In ellipsometric spectroscopy, an elliptically polarized light is allowed to reflect on the interface and the change in ellipticity and phase angle are determined from complex reflectivity.

$$\text{Complex Reflectivity} = \frac{^p \text{Reflectivity}}{^s \text{Reflectivity}}$$

$$\rho(E) = \frac{r_p\,(E)}{r_s\,(E)} = \tan\psi \cdot e^{(i\Delta)}$$

where ψ is ellipticity and Δ is phase change. These two parameters are related to the optical properties of the film.

$$\varepsilon_1 - i\varepsilon_2 = \varepsilon(a)\,\sin^2\theta\left(1 + \frac{1-\rho}{1+\rho}\right)\tan^2\theta$$

where θ is the angle of incidence

$$\varepsilon_1 = n - k$$

$$\varepsilon_2 = 2nk$$

where n and k are the refractive index and extinction coefficient respectively.

The electrode for the ellipsometric experiments was prepared by chemical vapor deposition of copper in argon atmosphere. Complex dielectric function of the copper surface was determined in air for the photon energy range of 400 to 800 nm (Figure 8). Complex dielectric function is low in the near IR region and starts to increase around 600 nm and reaches a maximum around 500 nm. Very little change was observed in the 400 - 500 nm region. Real dielectric function increases with light energy up to 550 nm and remains almost a constant in the high

After Development (Innerlayers) [LA8]

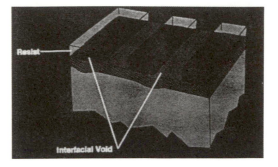

After Etching   [LA9]

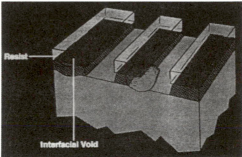

After Etching/Stripping   [LA10]

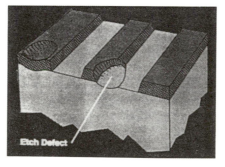

Figure 5: Conformance of electrochemically deposited films.

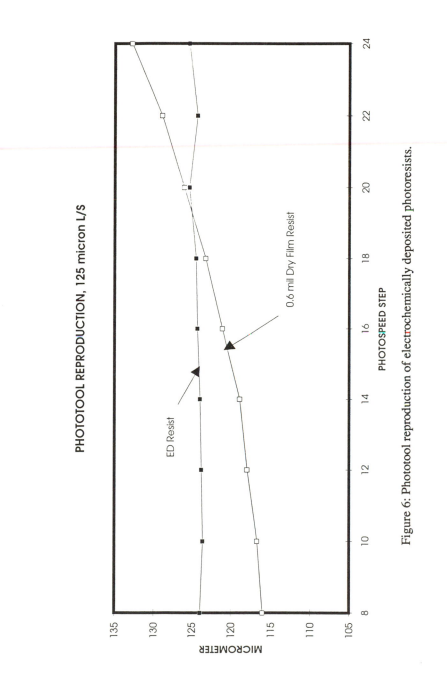

Figure 6: Phototool reproduction of electrochemically deposited photoresists.

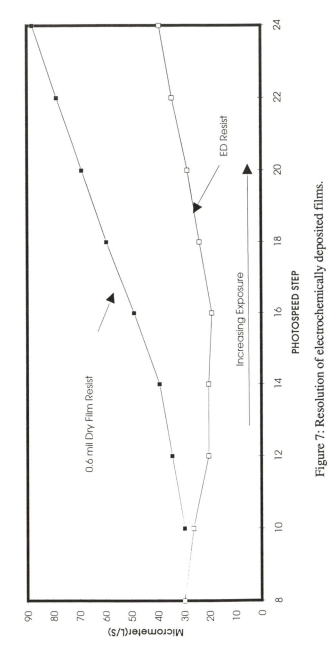

Figure 7: Resolution of electrochemically deposited films.

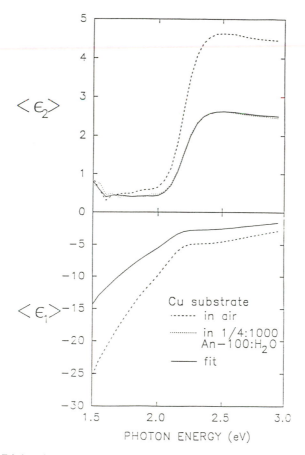

Figure 8: Dielectric functions of copper-air interface and copper-electrolyte interface.

energy region. From the complex dielectric function data refractive index and extinction coefficient can be determined as a function of light frequency. Diluted Primecoat AN-100 solution was carefully added to the cell without altering the configuration. Complex dielectric function was determined again under identical conditions (Figure 8). Complex dielectric function was found to vary significantly in the high photon energy region and the real component of the dielectric function varied more in the low photonic energy region. Using the refractive index and the extinction coefficient determined for the copper surface, refractive index and the extinction coefficient for the electrolyte solution were determined by linear regression analysis curve fitting (Figure 9). Refractive index for the electrolyte increases with the light frequency and the extinction coefficient decreases with frequency. The electrolyte absorbs only a small fraction of incident light in this region.

In the absence of any applied potential, the optical properties, namely the real and imaginary parts of the dielectric function of the copper, did not change over a period of 108 sec. This indicates that the electrode surface is stable under these conditions and no film growth occurs. A constant DC voltage of 20 V was applied, and the changes in the polarization and phase angle of the reflected light were monitored at a photon energy of 2.7 eV for 2500 sec. Changes in the real and imaginary part of the pseudo-dielectric function of the growing film/electrode combination were observed as soon as the electrical potential was applied (Figure 10). As soon as the voltage is applied, both real and imaginary components of the dielectric function changed. After an initial drop, the change in the real part of the dielectric function was marginal whereas the complex part of dielectric function continued to increase with time until the experiment was discontinued. A plot of real and imaginary components of the dielectric function is shown in Figure 11. Point A is the onset of the experiment. Both real and imaginary component remain constant in the beginning. As soon as the voltage is applied an increase in the imaginary component is observed. From point B to C change in real component is larger than the change in the imaginary component. The slope changes from point C to D. Very little change was observed in the dielectric functions with time after the point D. Dielectric functions of the copper surface and the electrolyte calculated from the above experiments were used to calculate the dielectric functions of the photopolymer film deposits on the copper surface. A polymer film and electrolyte at the interface were assumed. The changes in the dielectric function were fit to a structural model for film growth. Calculated dielectric functions for hemispherical islets of different spacing are shown in Figure 12. Initial nucleation of the polymer was consistant with hemispherical islands distributed on a square grid of 100 A spacing. The volume of water at the interface remained almost constant as the film continued to grow in thickness. This is attributed to the growth of the film in the vertical direction perpendicular to the electrode surface. The thickness of any material of bulk density is negligible under these conditions. Very little spreading occurs. The hemispherical islands tend to spread when the islets height approaches 50 A. This spreading continues until a smooth film is formed. Upon coalescence, bulk film formation commences and the film deposition occurs over the entire electrode surface. The film growth rate on

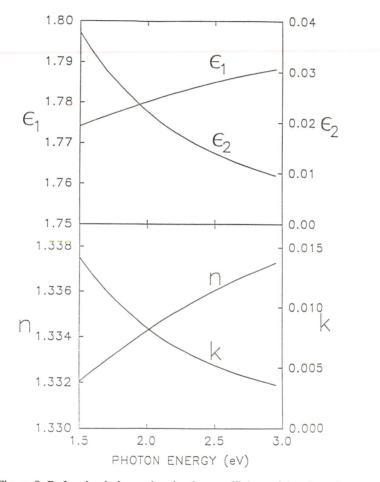

Figure 9: Refractive index and extinction coefficient of the electrolyte.

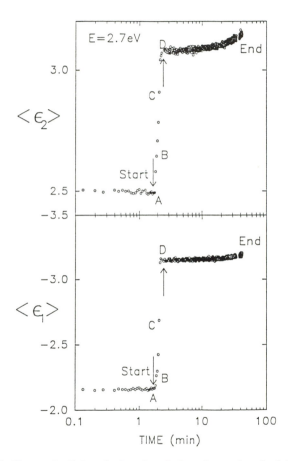

Figure 10: Change in dielectric function during electrochemical deposition.

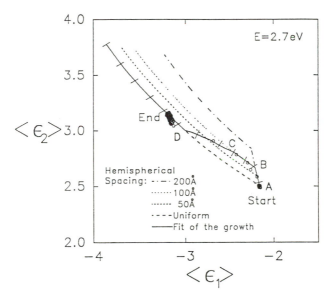

Figure 11: Hemispherical spacings of the islets formed during initiation deposition.

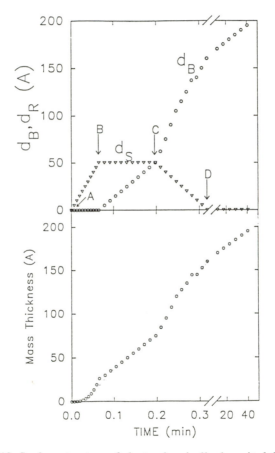

Figure 12: Surface structure of electrochemically deposited thin film.

the copper surface and on the polymer particles is similar for very thin film. During this bulk film growth, surface roughness is still present. The peak to valley difference is about 50 A. As the film thickness increases above 100 A, the surface roughness decreases slowly, and a smooth film is obtained when the film thickness is above 175 A. Additional deposition up to 195 A did not result in any significant change in the film properties. Film growth stopped when the DC potential was removed.

Surface structure of the electrode surface as a function of time is shown in Figure 12. From point A to B there is no bulk film thickness. Islets of 100 A spacings grow in vertical direction up to 50 A. After that the film tends to spread and the bulk film starts to grow on the entire surface. During this time the surface roughness remains constant. When the bulk film thickness reaches a height of about 50 A, surface roughness starts to decrease and a bulk film growth is seen all over the electrode.

The film growth rate was determined to be 550 A/minute initially. The electrodeposition rate decreases after the bulk film is formed. The film growth rate decreased by about 2 orders of magnitude after two minutes. Initial fast film growth is attributed to insoluble copper-carboxylic polymer complex formation(6). Once the copper electrode surface is completely covered with the photopolymer film, the deposition rate is controlled by the diffusion of copper ions or hydrogen ions through the film. Also electrochemical oxidation of water on the polymer surface might decrease the pH of the micro environment. In addition, conductivity of the metal-free film on the surface decreases leading to slower film deposition rate.

Dielectric functions of the electrode and the bulk film are shown in Figure 13. Dielectric functions of the film decrease with frequency in the low energy region. Complex dielectric function starts to increase after reaching a minimum at 2.5 eV. in the high energy region. The electrochemically deposited photopolymer thin film absorb strongly in the rear infra red region. Increase in absorbance in the near UV region is also observed. Extinction coefficients are similar to that of the electrochemically deposited thick film, which indicates that there is no other chemical reaction involved during the electrochemical deposition of thin film.

The thin film behaves like a free electron conductor. It is proposed that there are sufficient copper ions in the thin film to make it fairly conducting. As the film thickness increases, the conductivity of the film decreases at constant potential and consequently the deposition rate decreases. When the film thickness is above 15 microns there is practically no further deposition of the photopolymer film.

Conclusions

Time resolved ellipsometry is a surface sensitive technique that can be used to study the kinetics and mechanism of electrochemical deposition of photopolymer films.

Electrochemical deposition of photopolymer films occur on low energy surface states on the metal electrodes. The deposits tend to grow as "pillars" perpendicular to the electrode surface. A smooth coalesced film is observed when the thickness exceeds 175 A. The thin electrodeposited film is a good electrical conductor. The electrical conductivity decreases rapidly with thickness. Initial

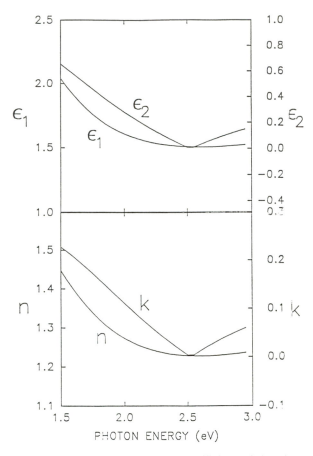

Figure 13: Refractive index and extinction coefficient of the electrochemically deposited thin film at the interface.

electrodeposition of the film is due to copper-complex formation. Protonation mechanism dominates when the electrode surface is covered with the photopolymer film.

References

1. Chandrasekaran, K; IS&T 9th International Congress Proceedings, **1993**, 263.
2. Beck, F; "Fundamental Aspects of Electrodeposition of Paint", in "Progress in Organic Coatings", **1976**, 4,1 Elsevier Sequiola S. A., Lasagna.
3. Collins, R. W; and Kim, Y; Anal. Chem., **1990**, 62, 1.
4. Kim, Y. T; Allara, D. L; Collins, R. W; Vedam, K; Thin Solid Films, **1990**, 193, 350.
5. Kenji Seko, Toshio Kondo, US Patent, 4,848,012 (1989).
6. Hamilton, W. L; Unpublished Results.

Chapter 14

Photopolymerization of Novel Degradable Networks for Orthopedic Applications

Kristi S. Anseth[1], Dina C. Svaldi[1], Cato T. Laurencin[2], and Robert Langer[3]

[1]Department of Chemical Engineering, University of Colorado, Campus Box 424, Boulder, CO 80309–0424
[2]Department of Orthopaedic Surgery, Medical College of Pennsylvania and Hahnemann University, Philadelphia, PA 19129
[3]Department of Chemical Engineering, E25–342, Massachusetts Institute of Technology, Cambridge, MA 02139

Multifunctional anhydride monomers (and oligomers) were developed that react to form highly crosslinked polyanhydride networks. Functionalized monomers (and oligomers) of sebacic acid and 1,6-bis(p-carboxyphenoxy)hexane were synthesized and subsequently photopolymerized. The high concentration of double bonds in the system and the multifunctional nature of the monomer (two double bonds per monomer molecule) led to the formation of a highly crosslinked polymer system in a period of seconds, depending on the initiation rate. Because the crosslinks were hydrolyzable anhydride linkages, the material remained biodegradable and the rate of degradation could be controlled by changes in the network composition and the crosslinking density. The high degree of crosslinking facilitated enhancement of mechanical properties (as compared to their linear counterparts) and also promoted a surface controlled degradation mechanism. The ability to photoinitiate the polymerization could lead to a new generation of orthopaedic implants that would provide surgeons with tremendous ease of placement and greater flexibility in design of fracture fixation systems.

Numerous musculoskeletal applications would benefit from recent advances in the development of safe, strong, easily processed, and degradable polymers. For example, the established standard for treatment of many fractures, especially those involving weight bearing bones, is the use of metal plates for fixation. However, the stiffness of metal plates is at least an order of magnitude higher than bone and can lead to stress-shielding problems where the bone can actually lose mass and become weak [1]. Metal plates also require a second surgery for removal and the stress-shielded bone is often prone to refracture.

Thus, the development of degradable polymeric implants would provide several advantages over metallic orthopaedic devices by eliminating the need for implant retrieval and circumventing problems associated with stress-shielding and bone atrophy. Degradable polymeric implants can also be used simultaneously to deliver therapeutic drugs or growth factors to enhance healing. An ideal polymer material for treating skeletal deficiencies should provide temporary mechanical strength to stabilize the osteotomy or fracture. Then, as the local region begins to reconstitute itself with new bone, the material should degrade (in a controlled fashion) into nontoxic molecules that the body can resorb or excrete. The degradation should proceed at a rate to allow transfer of stresses to the newly formed bone so that the usual process of remodeling occurs. Finally, consideration should be given to the intended clinical use of the material to minimize the difficulties in handling and placement for the surgeon.

To date, the most commonly used degradable polymer materials for fracture fixation, poly(lactic acid) (PLA), poly(glycolic acid) (PGA), and poly(p-dioxanon) (PDS), lack the versatility for extensive use in musculoskeletal applications. The limitation to broader application of these polymers, all polyesters, is linked to their bulk degradation mechanisms. Bulk degradation is a homogenous process which results in dramatic changes in molecular weight at short time scales (resulting in rapid loss of mechanical strength) and a majority of mass loss at the end of degradation (resulting in a local burst of acid products). In PLA/PGA implants, persistent abscess formation and lack of sufficient material strength for treating major long bone defects limit the polymers' use in many fracture fixation applications [2-8]. PDS has been successfully used for osteochondral fractures, but does not possess the mechanical properties for treating deficiencies in load-bearing bones [9-13].

In this contribution, polyanhydrides have been investigated as an alternative to polyesters for potential orthopaedic applications. Degradable polyanhydrides have been recently synthesized and shown to be safe in two phase III clinical trials, and a protocol for brain cancer treatment was approved by the FDA making them available to all patients in the U.S. [14]. Hence, from a clinical perspective, we have chosen to work with polymers that have a previous history indicating their safe application in humans. In contrast to the polyesters, polyanhydrides are surface degrading polymers (e.g., erode like a bar of soap), which is advantageous for drug delivery (prevents dose dumping) and load-bearing applications (maintains structural integrity). However, one limitation of existing polyanhydrides is their lack of initial mechanical strength required in load-bearing orthopaedic applications. One of the main objectives of this research, therefore, was to develop a high-strength, surface-eroding polyanhydride to meet the demands of orthopaedic fracture fixation devices.

In addition to developing high-strength polyanhydrides, photopolymerizations were also examined in an effort to develop a technique that would provide complete ease of fashioning and flexibility during implantation. Development of a photopolymerizable system would be beneficial for many reasons including fast curing rates at room temperature, spatial control of the polymerization, and complete ease of fashioning and flexibility during implantation. These benefits have led to advances in dentistry [15,16], ophthamology [17], and adhesion prevention [18], but currently no orthopaedic systems have been designed. Photoinitiation could greatly simplify the clinical application of orthopaedic polymer implants. For example in pin applications, a viscous, liquid monomer could be introduced into a pin hole and the system photopolymerized *in situ* to render a hardened polymer of the required dimensions. *In situ* polymerization eliminates the need for shaping the implant with blades, burrs, and warming instruments. Additionally, the process provides a faster and better mechanism for replicating complex shapes and should improve adhesion of the polymer implant to the bone.

In summary, this work presents a new class of photopolymerizable, multifunctional anhydride monomers (and/or oligomers) which react to form highly crosslinked polyanhydride networks. The presence of a high degree of crosslinking facilitates not only greatly improved mechanical properties, but also promotes a surface controlled degradation mechanism. Because the crosslinks are hydrolyzable anhydride linkages, the material remains biodegradable and the rate of degradation can be controlled by changes in the network composition and the crosslinking density.

Experimental

Materials. Sebacic acid (SA) and methacrylic anhydride (MA) were used as received from Aldrich, and 1,6-bis(carboxyphenoxy) hexane (CPH) was synthesized as described elsewhere [19]. Photopolymerizations were initiated with either ultraviolet or visible light. In the ultraviolet regime, three concentrations of a UV initiator Irgacure® 651 (I651, Ciba Geigy) (0.1 wt%, 0.5 wt% and 1.0 wt%) were investigated. The visible light initiating system consisted of camphorquinone (CQ, Aldrich) and ethyl-4-N,N-dimethylaminobenzoate (4EDMAB, Aldrich). Samples of monomer with three different concentrations of CQ/4EDMAB (0.2 wt%/0.8 wt%, 0.25 wt%/0.5 wt%, and 1.0 wt%/1.0 wt%) were polymerized using a visible light curing unit. This visible light initiating system is typical in many current dental applications [20].

Methods. Infrared spectroscopy (Nicolet Magna 550 FTIR), ^{1}H NMR spectroscopy (Nicholet 360 MHz), and gel permeation chromatography (Perkin-Elmer, isocratic LC pump 250, oven 101, and LC-30 RI detector at 254 nm) were used to characterize reaction products during the functionalization of the diacid monomers. Differential scanning photocalorimetry (DPC, Perkin-Elmer DSC7) and infrared spectroscopy were used to monitor the cure behavior of the crosslinking monomers. FTIR polymerization studies were conducted for both ultraviolet and visible light initiating systems. The total double bond conversion was calculated based on the decline in absorbance near 1640 cm^{-1}, the characteristic absorbance of methacrylate double bonds [21]. Samples were polymerized with either an ultraviolet light (EFOS, Ultracure 100SS) or a blue light (Curing Light XL1500, 3M Dental Products) curing system at various light intensities. Ultraviolet polymerizations were conducted with light intensities averaging between 1-150 mW/cm^2. Visible light polymerizations were performed at a light intensity of approximately 150 mW/cm^2. Mechanical properties of the resulting polymers were measured using a dynamic mechanical analyzer (Perkin-Elmer, DMA7) and degradation rates were characterized by mass loss.

Results and Discussion

Synthesis. Functionalized monomers (and oligomers) of sebacic acid (SA-Me$_2$) and 1,6 -bis(p-carboxyphenoxy)hexane (CPH-Me$_2$) were synthesized and subsequently photopolymerized as illustrated in Figure 1. First, the dicarboxylic acid was converted to an anhydride by heating at reflux in methacrylic anhydride for several hours. The dimethacrylated anhydride monomer was subsequently isolated and purified by dissolving in methylene chloride and precipitation with hexane. Infrared spectroscopy (IR), nuclear magnetic resonance (NMR) spectroscopy, and elemental analysis results indicated that both acid groups were converted to the anhydride, and the double bond of the methacrylate group was clearly evident.

Photopolymerizations were initiated with either ultraviolet or visible blue light of varying intensity (1-150 mW/cm^2). In general, the high concentration of double bonds in the system and the multifunctional nature of the monomer (two double bonds per monomer molecules) led to the formation of a highly crosslinked polymer system in a period of a few seconds, depending on the initiation rate.

In addition to the diacid molecules shown in Figure 1 (SA and CPH), several other diacid molecules are being explored and their structures are given in Figure 2. By developing an arsenal of functionalized monomers and oligomers, the flexibility to design polymers and copolymers that meet the various demands of materials properties (e.g., degradation rate, mechanical properties) for orthopaedic applications will be greatly enhanced. For example, as an alternative to sebacic acid, dodecanedioic acid can be used which is more hydrophobic. The degradation rate of the polymer should be strongly influenced by the overall hydrophobicity of the polymer. In addition, the first polyanhydrides studied for the delivery of drugs were based on bis(p-carboxyphenoxy)methane (CPM) [22], which is closely related to CPH. This class of monomers was selected for its hydrophobicity and a toxicological assessment that the monomers would be harmless and excretable [23]. By increasing the number of methylene groups from 1 (CPM) to 3 (CPP) to 6 (CPH), the hydrophobicity and degradation of the polymer can be further controlled. Additionally, amino acid containing polyanhydrides have been synthesized that contain imide linkages [24,25]. Imide bonds lead to higher glass transition temperatures and increased mechanical strength. Hence, by methacrylating and copolymerizing these various monomers over a wide range of compositions, we hope to span a broad range of degradation rates, physical properties, and mechanical properties.

Polymerization Behavior. Both Fourier-transform infrared spectroscopy (FTIR) and differential scanning photocalorimetry (DPC) were used to characterize the polymerization behavior, curing time, and maximum double bond conversion in these systems.

FTIR results are summarized in Figure 3. Specifically, the effects of initiation rate and the wavelength of the initiating light (i.e., 470-490 nm blue light or 365 nm UV light) on the curing time and double bond conversion were investigated. In developing systems for orthopaedic applications, both the polymerization time and maximum conversion are critical factors. In particular, the presence of unreacted

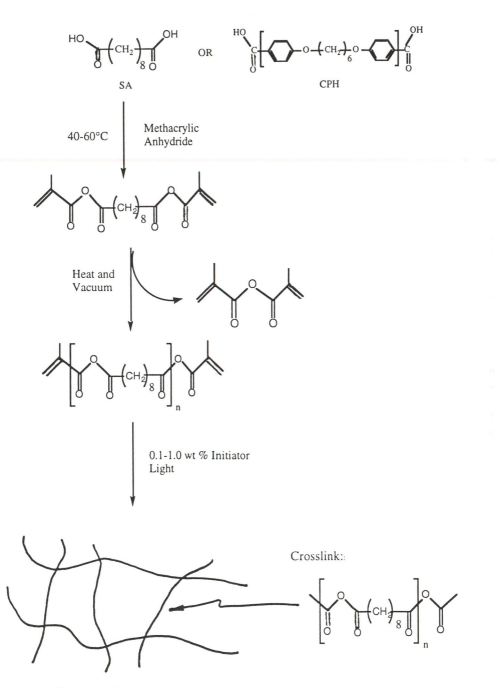

Figure 1. Polymer synthesis scheme for crosslinked polyanhydrides.

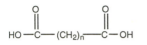

n=8 sebacic acid (SA)
n=10 dodecanedioic acid (DA)

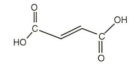

fumaric acid (FA)

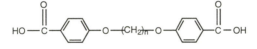

n=1 bis(*p*-carboxyphenoxy)methane (CPM)
n=3 1,3-bis(*p*-carboxyphenoxy)propane (CPP)
n=6 1,6-bis(*p*-carboxyphenoxy)hexane (CPH)

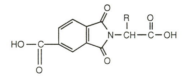

N-trimellitylimido acid

Figure 2. Structure of diacid molecules that can be functionalized to create crosslinked polyanhydride networks.

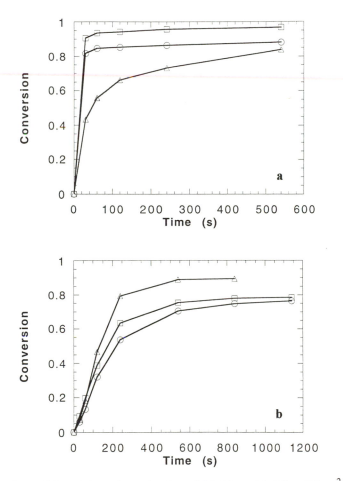

Figure 3. (a) Ultraviolet polymerization of SA-Me$_2$ with 150 mW/cm^2 of UV light and 1.0 wt% (☐), 0.5 wt% (○), and 0.1 wt% (△) I651. (b) Visible light polymerization of SA-Me$_2$ with 150 mW/cm^2 of blue light with 3 different concentrations of CQ/4EDMAB: 1.0 wt% /1.0 wt% (△), 0.25 wt%/0.5 wt% (☐), and 0.2 wt%/0.8 wt% (○).

monomer is significant and could adversely affect the mechanical integrity of the proposed polymer system. Hence, these studies were conducted to determine the clinical time necessary to cure the polymer and the maximum attainable double bond conversion.

In Figure 3 parts a and b, the effects of initiating light and initiator concentration on the polymerization rate of SA-Me$_2$. are shown. In general, higher light intensities or initiator concentrations enhance the rate of polymerization by increasing the rate at which active centers are generated. For example, in Figure 3a, after 30 seconds of exposure to approximately 20 mW/cm^2 of UV light, the system with 1.0 wt% initiator reached 91% conversion, whereas the system with 0.5 wt% initiator reached 82% conversion and the system with 0.1 wt% reached only 43% conversion. Hence, the clinical time for curing and setting of the polymer *in vivo* can be controlled by simple changes in the initiator concentration.

The polymerization rate was essentially zero in each of the systems (even with unreacted double bonds present and continued initiation) after 9 minutes of exposure to UV light. The maximum functional group conversion reached in each system was 96% (1 wt% I651), 87% (0.5 wt% I651), and 83% (0.1 wt% I651). If equal reactivity of the double bonds is assumed, only between 0.16 to 2.89% of unreacted monomer will be present at these total double bond conversions. Unreacted monomer can affectively plasticize the polymer network rendering it more pliable and decreasing its mechanical properties, and unreacted monomer may compromise the biocompatible nature of the system if the monomer leaches to a toxic level. Therefore, it is desirable to identify polymerization conditions which maximize the conversion of monomer.

Attainment of a maximum double bond conversion is typical in multifunctional monomer polymerizations and results from the severe restriction on bulk mobility of reacting species in highly crosslinked networks [26]. In particular, radicals become trapped or shielded within densely crosslinked regions known as microgels, and the rate of polymerization becomes diffusion limited. Further double bond conversion is almost impossible at this point, and the polymerization stops prior to 100% functional group conversion. In polymeric dental composites, which use multifunctional methacrylate monomers, final double bond conversions have been reported ranging anywhere from 55-75% [22,27-29].

In addition to the attainment of a maximum double bond conversion, Figure 3 also shows the dependence of the final conversion on the initiation rate. In general, faster rates of initiation/polymerization (e.g., higher initiator concentrations, higher light intensities) lead to greater double bond conversions. This increase in conversion is due to the effects of volume relaxation/physical aging on the system mobility. In essence, polymerization occurs at a rate that surpasses the rate of volume shrinkage and excess free volume is created. This excess free volume leads to higher mobility of reacting species, and thus, higher conversions. Compared to linear polymer systems, volume shrinkage plays an increasingly important role in the kinetics of multifunctional monomer polymerizations, particularly since these systems tend to gel at very low double bond conversions and relax very slowly to equilibrium. This effect has important clinical implications since higher conversion can be reached by polymerizing at higher light intensities, but the potential effect of internal stresses and physical aging on the polymer properties should be investigated.

Figure 3b contains FTIR results from the visible light polymerization of SA-Me$_2$ at varying concentrations of CQ/4EDMAB and 150 mW/cm^2 of blue light. The trends observed are similar to those seen during UV polymerization, but the rate of polymerization is significantly slower for this initiating system. The decreased polymerization rate was attributed to two factors. The primary factor is the efficiency of the CQ/4EDMAB initiatior system, which is considerably lower than the efficiency of the I651. The initiator efficiency is defined as the fraction of radicals formed in the primary step of initiator decomposition which are successful in initiating polymerization [30]. Side reactions and chain transfer to initiator decrease the initiation efficiency.

A second factor (which could potentially affect ultraviolet initiators as well) is the attenuation of light through the sample. Depending on the thickness of the sample, the molar absorptivity of the initiator (ε), and the concentration of the initiator ([A]), the differences between conversion at the surface and in the bulk of the sample can be appreciably different. These differences are the result of an exponential decay in the light intensity as a function of depth in the sample.

$$I = I_o \exp(-\varepsilon[A]x) \tag{1}$$

Here, I$_o$ is the incident light intensity, and I is the light intensity at a depth x. For the CQ/4EDMAB (0.2wt%/0.8wt%) initiating system, ε[A] $\approx$ 0.4 cm^{-1} between 470-490 nm (i.e., the wavelength of the polymerizing light). Under these conditions, I/I$_o$ is 0.82 at a sample depth of 0.5 cm, 0.67 at 1 cm, and 0.135 at 5 cm. Clearly the effect of the molar absorptivity and initiator concentration will be important parameters in the 3-D curing of these systems, and a model has been developed to relate the effects of light attenuation on the polymerization behavior and conversion [15,31].

Mechanical and Degradation Properties. Studies characterizing the mechanical properties of these highly crosslinked materials indicate properties that are intermediate between those of cortical and trabecular bone. Table I summarizes these results along with the mechanical properties of bone.

Table I. Selected mechanical properties of bone and the crosslinked polyanhydrides [32,33]

Material	Modulus		Strength	
	Tensile	Flexural	Tensile	Compressive
Cortical Bone	17-20 GPa	3 GPa	120-140 MPa	120-170 MPa
Trabecular Bone	50-100 MPa	50-100 MPa	1.2 MPa	1.9 MPa
Crosslinked SA	2 GPa	1.8 GPa	30-50 MPa	60 MPa
Crosslinked CPH	1.5 GPa	1.5 GPa	30-50 MPa	80 MPa

In comparison to linear polyanhydrides, the tensile modulus and strength of the crosslinked polyanhydrides are greatly improved. Other approaches to increasing the mechanical strength of linear polyanhydrides have focused on incorporating imide groups into the polymer backbone. These materials have good compressive strengths (between ~30-60 MPa) [24,25], but they are relatively brittle and lack tensile strength. Finally, currently used orthopaedic fracture fixation devices of poly (lactic acid) (PLA) have initial tensile strengths similar to the crosslinked polyanhydrides, between 30-50 MPa, which can be further increased by self-reinforcement techniques [34]. However, the efficacy of polyesters such as PLA and PGA in many orthopaedic applications is limited by their bulk degradation, which can lead to dramatic losses in mechanical strength early during the degradation process [34]. In contrast, Figure 4 shows the effect of mass degradation on the tensile modulus of the crosslinked polymers of SA and CPH. In both systems, the polymers maintain their structural integrity and greater than 90% of their tensile modulus at 40% mass degradation. Thus, the photopolymerizable and crosslinkable polyanhydrides provide not only great flexibility in placement and processing, but also enhanced mechanical properties of the resulting polymer.

Preliminary *In Vivo* Studies. A 3 mm model bone defect was created in the anteromedial tibial metaphysis of a Sprague-Dawley rat by drilling a 3 mm diameter by 5 mm hole through the near cortex and underlying trabecular bone using a high speed drill with a 2 mm carbide burr bit. A moldable putty comprised of SA-Me$_2$ and 0.5 wt% I651 was pressed into the defect and cured *in vivo* for 2 minutes using 10 mW/cm^2 of ultraviolet light. Macroscopic observations indicated that the polymer was well-adhered and filled the defect structure. The tibia was retrieved 7 days after surgery. Gross observations of the implant site indicated a ring of new bone growth that was approximately 1.0 mm thick, and the polymer was still intact at the core of the defect. The implant site along with the surrounding bone, was fixed in 10% neutral buffered formalin for histological analysis. The histology results are shown in Figure 5.

Conclusions

A new class of photopolymerizable, multifunctional anhydride monomers (and oligomers) was developed that react to form highly crosslinked polyanhydride networks. The reaction parameters such as the initiator concentration, wavelength of initiating light (blue versus ultraviolet), and intensity of light were investigated to determine the time necessary for clinical placement and the maximum functional group conversion. In the ultraviolet cured systems, greater than 90% conversion of the double bonds could be achieved in less than 60 seconds. In contrast, polymerization rates were slower with the visible light initiating system and was attributed to the lower initiation efficiency of this system. The high degree of crosslinking in these polymers facilitated enhancement of both compressive and tensile mechanical properties (as compared to their linear counterparts) and also promoted a surface controlled degradation mechanism. The mechanical properties were intermediate between those of trabecular and cortical bone. The ability to photoinitiate a polymerization *in vivo* could lead to a new generation of orthopaedic

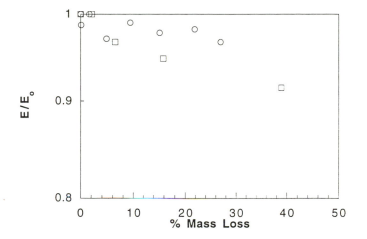

Figure 4. The tensile modulus normalized by the initial tensile modulus during the degradation of crosslinked SA ($\square$, E_0=2 GPa) and crosslinked CPH ($\bigcirc$, E_0=1.5 GPa) at 37°C in phosphate buffer solution (pH=7.4).

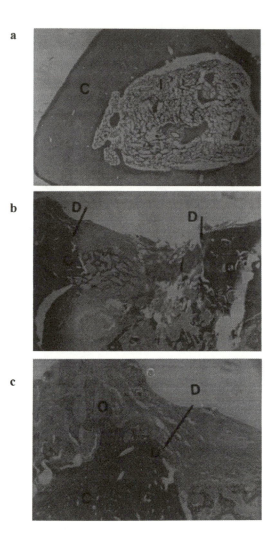

Figure 5. 3 mm defect study in a rat tibia showing a control (no defect) and a partially healed defect site. Pictures are transverse sections through the tibia and defect.

a) Control sample showing normal bone morphology with no drilled defect. C -- undisrupted cortical bone. I -- intramedullary canal.

b) View showing the drill entry site and edges of the defect in the cortical bone. C -- cortical bone. D -- edge of drilled defect.

c) Enlarged view of the defect edge showing morphology of the new bone growth. C -- cortical bone. D -- edge of drilled defect. O -- new bone growth.

implants that would provide surgeons with tremendous ease of placement and greater flexibility in the design of fracture fixation systems.

Acknowledgments - The authors would like to thank the National Institutes of Health for its support of this work through a fellowship to KSA and NIH Grant No. AR41972.

Literature Cited

1. C.T. Brighton, G. Friedlaender and J.M. Lane, Bone formation and repair. AAOS (1994).
2. P. Rokkanen, S. Vainionpaa, K Vihtonen, M. Mero, H. Patiala, J. Kilpikari and P. Tormala, *Acta Orthop. Scand.*, *57*, 237 (1986).
3. P. Rokkanen, P. Axelson, J. Raiha, K. Sittnikow, K. Skutnabb, M. Mero, S. Vainionpaa and P. Tormala, *Acta Vet. Scand.*, *29*, 469 (1988)
4. P. Rokkanen, O. Bostman, A. Makela and P. Tormala, *J. Bone Joint Surg.*, *71*, 706 (1989)
5. S. Vainionpaa, K. Vihtonen, M. Mero, H. Patiala, P. Rokkanen, J. Kilpikari and P. Tormala, *Arch. Orthop. Trauma Surg.*, *106*, 1 (1986)
6. O. Bostman, U. Paivarinta, E. Partio, M. Manninen, J. Vasenius, A. Majola and P. Rokkanen, *Clin. Orthop.*, *285*, 263 (1992).
7. Y. Matsusue, T. Yamamuro, S. Yoshii, M. Oka, Y. Ikada, S. Hyon and Y. Shikinami, *J. Appl. Biomater.*, 2, 1 (1991).
8. R.W. Bucholz, S. Henry and M.B. Henley*J. Bone Joint Surg. Am.*, *76*, 319 (1994).
9. H. Greve and J. Holste, *Akt. Traumatol.*, *15*, 145 (1985).
10. L. Claes, C. Burri, H. Kiefer and W. Mutschler, *Akt. Traumat.*, *16*, 74 (1986).
11. W. Thomas, C. Lutten and H. Lorenz, *Sportverletz Sportschaden*, 2, 61 (1989).
12. W. Ruf, W. Schultz and K. Buhl, *Unfallchirurgie, 16*, 202 (1990).
13. G. Patton, M. Shaffer and D. Kostakos, *J. of Foot Surg.*, *29*, 122 (1990).
14. H. Brem, *et al., Lancet, 345*, 1008 (1995).
15. K.S. Anseth, S.M. Newman and C.N. Bowman, *Adv. Polym. Sci.* 122, 177 (1995).
16. J. Engelbrecht, *US Patent 4,872,936* (1989).
17. R.H. Grubbs, R.J. Coots and S.H. Pine, *US Patent 4,919,151* (1990).
18. J.L. Hill-West *et al., Obstet. Gynecol. 83*, 59 (1994).
19. A. Conix, *Macromol. Synth. 2*, 95 (1966).
20. I.E. Ruyter and H. Oysaed, *CRC Critical Reviews in Biocompatibility, 4*, 247 (1988).
21. K.S. Anseth, C. Decker and C.N. Bowman, "Real-time infrared characterization of reaction diffusion during multifunctional monomer polymerizations,"*Macromolecules, 28*, 4040-43 (1995).
22. H.B. Rosen, J. Chang, G.E. Wnek, R.J. Linhardt and R. Langer, *Biomaterials, 4*, 131 (1983).
23. J. Tamada and R. Langer, *J. Biomater. Sci. Polymer Edn., 3*, 315 (1992).
24. A. Staubli, E. Ron and R. Langer, *J. Am. Chem. Soc., 112*, 4419 (1990).
25. K.E. Uhrich, A. Gupta, T.T. Thomas, C.T. Laurencin and R. Langer, *Macromolecules, 28,*. 2184 (1995).
26. J.G. Kloosterboer, *Adv. Polym. Sci., 84*, 1 (1988).

27. J.L Ferracane and E.H. Greener, *J. Biomed. Mater. Res., 20*, 121 (1986).
28. D.J. Barron, R.A. Rueggeberg and G.S. Schuster, *Dent. Mater., 8*, 274 (1992).
29. K. Tanaka, M. Taira, H. Shintani, K. Wakasa and M. Yamaki, *J. of Oral Rehabilitation, 18*, 353 (1991).
30. G. Odian, *Principles of Polymerization*, McGraw-Hill: New York, 1982.
31. K.S. Anseth and C.N. Bowman, *Polym. Reaction Eng., 1*, 499 (1992-93).
32. M.J. Yaszemski, PhD Thesis, Massachusetts Institute of Technology, 1995.
33. Y.C. Fund, Biomechanics: Mechanical Properties of Living Tissues, 2nd Ed., Spring-Verlag, NY (1993).
34. P. Tormala, P. Rokkanen, J. Laiho, M. Taraminmaki and S. Vainionpaa, "Material for osteosynthesis devices," U.S. Pat. No. 4,743,257 (1988).

Chapter 15

Photopolymerizations of Vinyl Ester: Glass Fiber Composites

L. S. Coons, B. Rangarajan, D. Godshall, and Alec B. Scranton[1]

Department of Chemical Engineering, Michigan State University, A202 Engineering Building, East Lansing, MI 48824–1226

The spatial and temporal control afforded by photopolymerizations make them attractive for rapid and inexpensive processing of polymeric composites. We have investigated the production of thick polymers and glass fiber composites using commercially available vinyl ester resins. The studies illustrate the effects of light intensity, initiator formulation and concentration, exposure time, and fiber loading on the mechanical properties of the photocured composite materials. With moderate incident light intensities and high fiber loadings the photopolymerizations were found to proceed to completion in minutes to yield composites good mechanical properties. A hybrid photo/thermal initiation strategy in which heat released from an exothermic photopolymerization leads to the additional dissociation of a thermal initiator was found to be especially promising for the production of thick polymers and composites.

Photopolymerizations offer many advantages that may be exploited in the design of industrial processes. For example, photopolymerizable formulations are typically solvent-free, thereby minimizing volatile organic emissions. Moreover, since active centers may be produced rapidly and efficiently using photochemical processes, photopolymerizations provide high production rates, and are very energy efficient compared to thermal systems in which the entire reaction systems is raised to elevated temperatures. Finally, photopolymerizations provide a great deal of control over the initiation reaction: spatial control, since the light may be directed to locations of interest in the system; and temporal control, since the light may be readily shuttered on or off. This set of advantages has lead to tremendous growth in a host of photopolymerization applications including: films, inks, and coatings for a variety of substrates including paper, metal, plastic, and wood; dental materials for

[1]Corresponding author

which the initiating light is routed into the mouth using optical fibers; and a variety of high-tech and electronic applications such as coatings on optical fibers, replication of optical disks, and fabrication of printed circuit boards.

In most current applications of photopolymerizations the reactions are carried out in thin-film geometries using neat (unfilled) monomer. The advantages offered by these reactions also make them attractive for the development of new processes for producing fiber-filled polymer composites. Most notably, the spatial and temporal control of initiation offered by photopolymerizations provides "cure on demand" which could be used to achieve short cycle times while circumventing many problems and limitations of current composites production methods. In this paper we will focus on vinyl ester - glass fiber composites used in the durable goods industry. These composites are typically produced by resin transfer molding (1) or filament winding (2) processes in which a free radical polymerization is triggered using thermal initiators. In these thermally initiated processes, the time required for heat transfer places a lower limit on the cycle time, and attempts to reduce the cycle time by preheating the mold (in resin transfer molding) or the resin (in filament winding) invariably lead to problems associated with premature cure, or very complicated flow patterns.

As an example of how the cure on demand afforded by photopolymerizations could lead to improvements in composites manufacturing, consider the resin transfer molding (RTM) process. In this method, a reacting resin is injected into a closed mold containing the fiber preform. Short cycle times necessitate a careful balance between the conflicting requirements of rapid mold filling and high polymerization rates. This trade-off leads to a very complicated flow process in which a reacting liquid is forced through a porous medium (the preform). As the liquid reacts it becomes more viscous (actually viscoelastic), leading to several important problems associated with high operating pressures, poor resin impregnation both into and around the fibers, preform displacement, and possible gelation of the resin in the transfer lines. These problems, which ultimately arise from the fact that the mold-filling and reaction are coupled to one another, are exacerbated by attempts to enhance the cycle time by using more reactive resins.

The temporal control offered by photopolymerizations could be exploited to alleviate many of the problems associated with RTM by: i) allowing the mold to be completely filled with a low-viscosity resin before any reactions takes place, and ii) initiating a fast thermoset reaction after the mold is completely filled. This decoupling of mold filling and reaction would permit complete control of the time allowed for micro-flow, and would reduce waste and expensive clean-up by preventing premature reaction (gelation in the transfer lines). In addition, unlike traditional processes, the photoinitiated reactions are not limited by the rate of heat transfer from the mold to the resin, and would consequently offer short cycle times. Finally, since there would be no need for high pressures or external heating of the die, the proposed process would be very energy efficient. Other composite processing methods such as filament winding would benefit in a similar manner from the cure on demand afforded by photopolymerizations.

Photopolymerizations of thick and fiber-filled polymers are more challenging than polymerizations of thin-film systems due to the exponential reduction in light intensity through the sample resulting from absorption. Therefore for thick polymers and composites, proper selection of the initiator formulation and illumination wavelength are imperative to ensure that the samples cure throughout (including the portions of the sample furthest from the light). In this contribution we will examine two strategies for polymerization of thick polymers and composites using UV light as the sole (external) initiating source.

The first method is based upon careful selection of the reaction system (monomer, fibers, photoinitiator, and initiating wavelength) such that the resin and fibers are transparent to the initiating wavelength, and only the initiator absorbs light. This criterion assures that any light intensity gradient initially present arises only from absorbance by the initiator. Secondly, the initiator must "photobleach" at the initiating wavelength (i.e. the initiating fragments should not absorb light at the initiating wavelength). Therefore the initiator becomes transparent upon production of active centers. Meeting these two simple criteria ensures that the photochemical event which leads to the initial light intensity gradient also leads to its elimination, thereby allowing light to penetrate deep into the sample. In this way, the "self-eliminating light intensity gradients" (*3*) method allows photopolymerization of thick polymer samples.

In the second dual photo/thermal initiation strategy, the approach described above is augmented by the inclusion of a thermal initiator. Upon illumination, active centers produced by fragmentation of the photoinitiator start the polymerization reaction. The heat evolved from the exothermic photopolymerization elevates the temperature of the system and results in the production of additional active sites by the thermal initiator. This dual initiating strategy provides both the cure on demand (temporal control) afforded by photopolymerization, and the completeness of cure provided by the thermal initiator.

Photopolymerizations of glass fiber composites are beginning to receive some attention in the literature. For example, in a series of papers, (*4-6*) Ogale and collaborators examined photopolymerizations of acrylate - glass fiber composites designed for three dimensional stereolithography applications. These investigators examined the mechanical properties of photocured composites containing continuous (*4*) or chopped (*5*) e-glass fibers, as well as 55 μm diameter borosilicate glass microspheres (*5*). They concluded that the photolithography of composites was indeed feasible, and that, for example, a photocured composite containing 15 vol.% continuous fibers exhibited a four-fold increase in tensile strength compared to the neat acrylate. In a third paper, (*6*) Renault *et al.* used dynamic mechanical analysis to examine the effect of glass, quartz, and carbon fibers on the photopolymerization rate of acrylate composites. The authors reported that the glass and the quartz fibers had no effect on the cure rate, while the carbon fibers reduced the rate of photopolymerization (*6*). In a previous contribution, (*3*) we reported that photocured composites up to one centimeter in thickness exhibited mechanical properties equivalent to their thermally cured counterparts.

In this contribution we present a series of studies on photoinitiated polymerizations of vinyl ester - glass fiber composites. The objectives of these investigations are i) to demonstrate the feasibility of photopolymerizations of thick and fiber-filled polymers based upon self-eliminating light intensity gradient and hybrid photo/thermal cure strategies using common commercially-available initiators and resins, and ii) to identify the important operational variables that may be used influence the cure rates and mechanical properties of the resulting polymers. Therefore, we have examined the relationship between reaction variables such as the initiator formulation and concentration, the incident light intensity and wavelength, the fiber loading, and the temperature on the reaction rate (cure time) and mechanical properties of the resulting composites.

Materials and Methods

One objective of this study was to demonstrate the feasibility of photopolymerizations of thick and fiber-filled polymers using the same constituents that are currently polymerized using thermal initiators. Therefore, photopolymerization studies were performed using a common vinyl ester resin (Derakane 470-45, Dow Chemical Company) and benzoin ethyl ether (BEE, Aldrich Chemical Company) as the photoinitiator. The fibers used in these studies were finely chopped E-glass fibers with very short aspects ratios. While these fibers do not maximize the mechanical properties of the composite (longer fibers would provide much more reinforcement) they were selected because they maximize the number of fiber - resin interfaces for a given fiber loading. Therefore, if scattering does not present a problem with this system, it will be even less of a problem with longer fibers. All materials were used as received. A 1000 watt Hg(Xe) arc lamp (model 6293, Oriel Corporation) was used as the source of initiating light. Light intensities were measured using a UVICURE Plus high energy UV integrating radiometer (over the 320-390 nm range). The incident intensity was varied by adjusting the condenser lens on the lamp or by altering the distance between the sample and the light source.

Photopolymerizable mixtures were prepared by adding a predetermined amount (typically 0.1wt%) of the initiator to the resin at room temperature while stirring on a magnetic plate. The fully formulated mixtures were stable in the absence of initiating light, and were stored in the dark at room temperature prior to use. Specimens for mechanical testing were prepared by carrying out the photopolymerizations in cylindrical glass molds irradiated radially. The powdered glass fiber was mixed with the formulated resin to form a slurry with a desired fiber loading. This slurry was poured into the mold and irradiated with unfiltered UV light for a predetermined length of time. The fully cured cylindrical composite specimens were ~8 cm long and 0.58 cm in diameter.

In situ temperature profiles during the photopolymerizations were obtained using a high-speed thermocouple system (WB-AAI-B8 interface card, Omega Systems) which was interfaced with a personal computer. For these studies, the polymerizations were carried out in cylindrical polyethylene molds with an inner diameter of 1.4 cm. These samples were irradiated axially from the top, and the molds were insulated with polyurethane foam to minimize heat loss from the sample. Each experiment was carried out using 1.0 ml of fully formulated mixture containing resin, glass fiber, and initiator. Type K thermocouples were positioned at the bottom and center of each sample. Temperature data were collected every 0.5 seconds for a total reaction time of twelve minutes. For these experiments, the 1000 W Hg(Xe) lamp was outfitted with a water filter to eliminate the incident infrared irradiation.

Ultraviolet-visible absorbance spectra were collected using a Hewlett-Packard UV-Visible Spectrometer model 8452A. The absorbance of the initiators and resins were obtained from dilute solutions of these compounds in 1-propanol. To characterize the photobleaching of benzoin ethyl ether at 328 nm, absorbance decay experiments were conducted using 5 ml quartz cuvettes containing solutions of BEE (0.1, 0.5, or 1.0 wt%) in 1-propanol. In these studies, each cuvette containing 3ml of the initiator solution was illuminated from the top under the Hg(Xe) arc lamp for a predetermined time interval, was removed from the light, mixed to eliminate any concentration gradients, then placed in the spectrometer to collect an absorbance spectrum. The low light intensity studies were performed using a 200 Watt Hg(Xe) arc lamp. For these studies, both a water filter and a 10 nm bandpass filter centered at 350 nm were used to filter out all but the near UV wavelengths.

The mechanical properties of composite specimens were determined using a United STM-20 instrument in accordance with the ASTM D 790 method. The flexural strength and flexural modulus were measured using the three point flexural test with a span length of 5.08 cm (2 inches), a 454 kg (1000 pound) load cell, and a downdrive rate of 0.0212 mm/sec (0.05 inches per minute). The flexural modulus was calculated for these samples at elongations between 0.0013% to 0.0064% of a meter. Average values of measurements from at least four different samples were used for each specimen, with the error bars representing high and low values.

Results and Discussion

Initiator Absorbance Studies. A series of studies were performed to characterize the absorbance spectra of the vinyl ester (Derakane) resin, the glass fibers, the benzoin ethyl ether (BEE) initiator, and the BEE initiator fragments. These absorbance spectra were used to identify an appropriate initiating wavelength since it is desirable to select a wavelength for which the resin and fibers and transparent, and only the initiator absorbs. The resin exhibited an absorbance maximum at a wavelength of about 260 nm, with diminishing absorbance up to about 310 nm. Similarly, the glass fibers exhibited an absorbance peak in the deep UV region, but were transparent to wavelengths above about 320 nm. These studies indicate that the near-UV region of the spectrum between about 320 and 400 nm is appropriate for initiation since it represents a clean spectral window for which the resin and fibers do not absorb.

Figure 1 contains absorbance spectra of BEE before and after 30 minutes of illumination at an intensity of 33 mW/cm^2. The figure illustrates that the absorbance peak between 300 and 390 nm decreases with illumination, suggesting that the absorbance of the initiator fragments in this region is considerably lower than that of the unreacted initiator. Experiments performed in the presence of an efficient hydrogen donor (tributylamine) exhibited much less photobleaching. Therefore, we attribute this loss of absorbance between 300 and 390 nm to a disruption of the π electron structure of the BEE upon photofragmentation. These results suggest that for the polymerization of thick polymers and composites, it is important to ensure that the active centers do not absorb the initiation light, which is more likely if these centers are produced by photofragmentation rather than hydrogen abstraction.

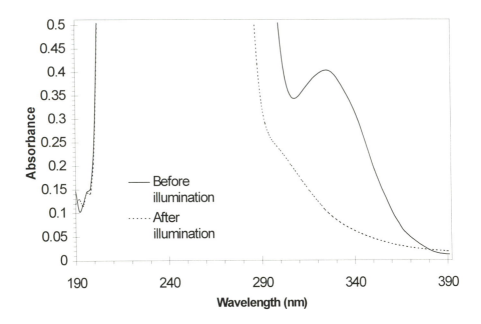

Figure 1. Absorbance spectra of benzoin ethyl ether before and after 30 minutes of illumination illustrating the absorbance decrease at wavelengths from 300 to 400 nm.

The simple absorbance experiments described above suggest that our reaction system meets the "self-eliminating light intensity gradient" criteria outlined in the introduction. Specifically, the resin and fibers are transparent to initiating wavelengths above 320 nm, and the initiator exhibits effective photobleaching at these wavelengths. Therefore, the photochemical event which leads to the initial light intensity gradient (absorption and production of active centers) should also lead to its elimination, thereby enabling light to penetrate deep into the sample. To investigate this self-eliminating behavior, a series of studies was performed in which the absorbance of BEE at 328 nm was monitored as a function of exposure time. As described in the *Materials and Methods* section, these studies were performed in 1-propanol solutions, and the 3 ml samples were well mixed before the absorbance was measured. Therefore the resulting absorbance profiles illustrate the effect of the photobleaching on the average absorbance in the sample.

Figure 2 illustrates the effect of light intensity on the average absorbance of the samples. First of all, this figure illustrates that for each light intensity the average absorbance decays exponentially with time as the initiator is consumed by photofragmentation and active centers are formed. Secondly, the figure illustrates that the decrease in the average absorbance occurs more rapidly as the incident light intensity is increased. To allow the absorbance decay profiles to be completely characterized, these studies were performed using very low light intensities from a 200 Watt lamp. Other studies performed at higher incident intensities (26, 36 mW/cm^2) revealed that the absorbance decay was finished in a few minutes. Therefore, these results indicate that even in samples a few centimeters thick, the decrease in absorbance occurs rapidly enough to allow polymers and composites to be produced in reasonable cycle times.

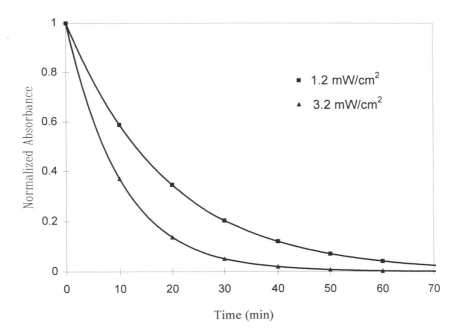

Figure 2.   Absorbance decay of benzoin ethyl ether at 328 nm as a function of illumination time. The experiments were performed on 3.0 ml solutions of 0.1 wt. % BEE in 1-propanol.

Figure 3 illustrates the effect of the initiator concentration on the absorbance decay rate. These studies were also performed at a low light intensity to allow the profile to be completely characterized (again, higher intensities provide much more rapid absorbance decays). The figure illustrates that, as expected, the absorbance decay rate decreases as the initiator concentration is increased. Again the absorbance decay follows an exponential decay profile, and the decay rate can be characterized by the exponential time constant (τ in the exponential decay equation $e^{-t/\tau}$). Figure 4 contains a plot of the exponential time constant as a function of the initiator concentration at constant incident light intensity. The figure illustrates that the exponential decay time increases linearly as the initiator concentration is increased. These absorbance decay studies illustrate that the initiator concentration and incident light intensity provide two process variables which may be used to achieve the desired cycle time in photopolymerizations of composites.

Temperature Profiles of Adiabatic Polymerizations. Experiments were conducted to characterize the adiabatic temperature profiles of photocured composites

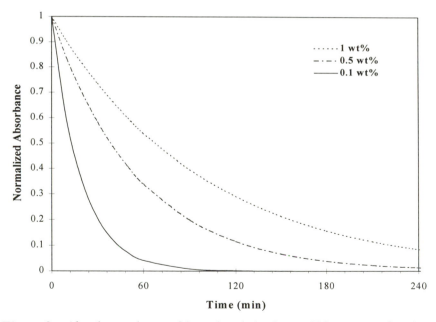

Figure 3. Absorbance decay of benzoin ethyl ether at 328 nm as a function of illumination time. The experiments were performed on 3.0 ml solutions of BEE in 1-propanol with an incident intensity of 1.2 mW/cm².

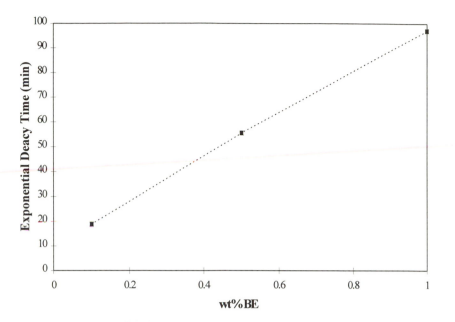

Figure 4. Exponential decay time constant for the absorbance decay of benzoin ethyl ether at 328 nm as a function of BEE concentration.

containing between 0 and 60 wt.% fibers. As described in the *Materials and Methods* section, these studies were performed using insulated polyethylene molds irradiated axially with the 1000 Watt lamp. These studies are important for establishing the feasibility of using light (with no external heating) as the sole source of initiating energy. The attainment of elevated temperatures during the reaction is important to achieve complete cure and full property development since a sample which is cured entirely at room temperature would exhibit a relatively low limiting conversion and a low glass transition temperature due to premature vitrification. Therefore, it is important to characterize the reaction temperature to determine whether the exothermic heat of polymerization leads to appropriate temperature profiles without external heating or cooling.

Figure 5 contains experimental profiles of the reaction temperature at the bottom of the sample as a function of time for nearly adiabatic photopolymerizations of Derakane resins containing between 0 and 60 wt.% of the glass fibers. The figure illustrates that for all fiber loadings, upon illumination the temperature exhibits an initial increase from room temperature to a final plateau value around 130°C. Moreover, the figure illustrates that as the fiber loading is increased, both the rate of the initial temperature increase, and the final plateau value, are reduced. These trends are easily explained by the reduction in the reactive fraction of the sample with increasing fiber loading. The important practical result of this study is that for all fiber loadings, the adiabatic reaction temperature is high enough to ensure adequate polymerization and property development of the composites, but low enough to prevent thermal degradation. These results suggest that with proper system design, it is feasible to perform photopolymerizations of composites with no external heating or cooling.

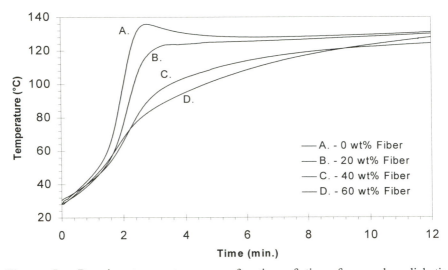

Figure 5. Reaction temperature as a function of time for nearly adiabatic photopolymerizations of vinyl esters with various fiber loadings.

Development of Mechanical Properties. Figure 6 contains a plot of the flexural modulus as a function of the cure time for composites prepared with fiber loadings ranging from 0 to 60 wt.%. These samples were polymerized in glass molds illuminated radially with an incident light intensity of 480 mW/cm^2. Each data point in the figure represents an average of at least 4 independent measurements, with the high and low values being depicted by error bars. Therefore, the error bars indicate the entire range of the experimental data rather than the standard deviation. Figure 6 illustrates two important results. First of all, even at this relatively low light intensity, the photopolymerizations proceed to completion in a reasonably short time. For example, even the sample with a 60% fiber loading is finished curing in approximately seven minutes (as indicated by the modulus reaching a peak). Secondly, as expected, the figure reveals that increased fiber loading leads to an increase in the modulus of the fully cured composite. For example, the flexural modulus increases by a factor of 3 at 50 wt% fiber loading and a factor of more than 4 at 60 wt.% fiber loading when compared to the unfilled sample.

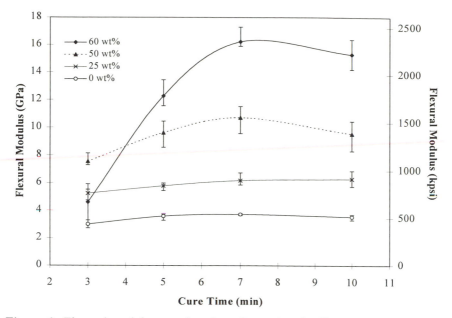

Figure 6. Flexural modulus as a function of cure time for fiber loadings between 0 and 60 wt.% with an incident light intensity of 480 mW/cm^2.

Figure 7 contains a plot of the flexural modulus after five minutes of illumination as a function of the initiator weight fraction. These studies were performed using neat (unfilled) resins exposed to incident light with an intensity of 480 mW/cm^2. Again, each data point in the figure represents an average of at least 4 independent measurements, with the high and low values being depicted by error bars. As shown in the figure, at low initiator concentrations the flexural modulus increases with increasing initiator concentration, but as the initiator level is increased above about 0.2 wt.%, the flexural modulus decreases with increasing initiator concentration. The figure illustrates that there is an optimum initiator concentration for achieving maximum strength in a given cure time. The decreasing modulus could arise from either i) optical density effects which prevent penetration of light into the sample and therefore lead to non-uniform or incomplete cure, or ii) a reduction in the primary chain length due to increased termination associated with the higher active center concentration. The fact that samples containing 2 and 4 wt.% initiator exhibited only a surface cure in this five minute time frame supports the first hypothesis. These results suggest that for photopolymerizations of thick polymers and composites there is a balance between having enough initiator to ensure adequate active center production, and having too much (which prolongs the time required for elimination of the light intensity gradient). Therefore there exists an optimal initiator concentration, which depends on factors such as initiator type, temperature, and light intensity.

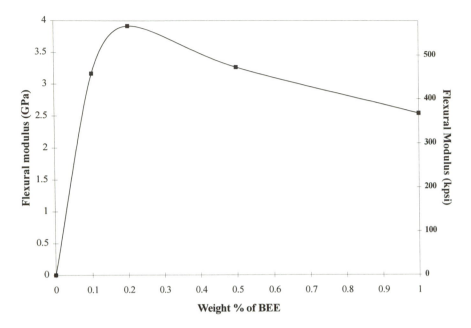

Figure 7. Flexural modulus after 5 minutes of illumination as a function of the initiator weight fraction for polymerizations of unfilled samples with an incident light intensity of 480 mW/cm^2.

Dual Photo/Thermal Initiation Studies. A series of studies were performed using reactive formulations containing both a photoinitiator and a thermal initiator dissolved in the Derakane resin. The objective of these studies was to investigate a dual cure strategy in which the heat liberated by the photo-induced polymerization leads to the production of additional active centers by the dissociation of a thermal initiator. In this way, the dual cure strategy could offer both the temporal control of the start of the reaction afforded by the photopolymerization, as well as enhanced reaction rate and completeness of cure provided by the thermal initiation.

The effect of the addition of the thermal initiator on the cure rate for a relatively low incident light intensity (52 mW/cm^2) is illustrated in Figure 8. This figure contains a plot of the time required for complete cure as a function of thickness for neat resin systems containing 0.1 wt.% BEE and 0, 0.2, and 1.0 wt.% benzoyl peroxide (BPO, a thermal initiator). The figure illustrates that samples containing only the photoinitiator exhibit cure times that increase linearly with thickness from approximately 8 minutes for a thickness of 2 mm, to 25 minutes for a thickness of 1.5 centimeters. In this case the time required for complete cure is controlled largely by the time required for elimination of the light intensity gradient. In contrast, the samples containing both the photoinitiator and the thermal initiator exhibit a cure time of approximately five minutes, nearly independent of thickness.

It is interesting that there is little change in the time required to cure the polymer as the concentration of BP is increased from 0.2 to 1.0 wt.%. These results suggest that the reaction proceeds to completion even before the light penetrates deep into the sample, perhaps by the propagation of a thermal front created by the initial photopolymerization at the leading surface. In any case, these results illustrate that the addition of a thermal initiator drastically reduces the polymerization time for thick systems.

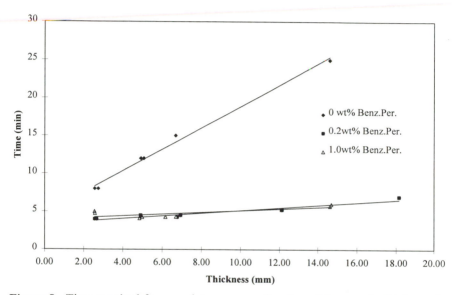

Figure 8. Time required for complete cure as a function of the sample thickness for polymerization systems containing both BEE and BPO.

Figure 9 contains a plot of the time required for complete cure as a function of thickness for neat resin systems containing 0.1 wt.% BEE and 0, 0.2, and 1.0 wt.% BPO with an incident light intensity of 900 mW/cm². The trend in Figure 9 is similar to that in Figure 8 since the sample containing only the photoinitiator exhibits a cure time that increases nearly linearly with thickness, and the addition of the photoinitiator results in both a reduction in the cure time, and a more gradual dependence upon thickness. Comparison of Figures 9 and 10 reveals that the increase in the incident light intensity has a significant effect on the cure rate. The increase in light intensity from 52 mW/cm² to 900 mW/cm² leads to a four-fold reduction in the cure time of the system containing only the photoinitiator. In contrast, the cure times of the hybrid photo/thermal systems are reduced by a factor of two for the same increase in incident light intensity. This difference, which may be attributed to the relative rates of photoinitiation, thermal initiation, propagation, and termination, will be addressed in detail in a future publication. For composites manufacturing, it is significant that the hybrid photo/thermal systems are completely cured in only a few minutes regardless of the sample thickness.

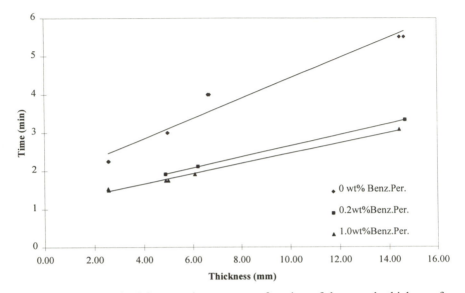

Figure 9. Time required for complete cure as a function of the sample thickness for polymerization systems containing both BEE and benzoyl peroxide.

Figure 10 illustrates that the dual photo/thermal initiation strategy is effective for polymerization of glass-filled composites. This plot contains a plot of the flexural modulus as a function of illumination time for Derakane resin systems containing 0.1 wt.% BEE and 0.2 wt.% BPO with fiber loadings of 0, 25, and 50 wt.%. Again, the data points in the figure correspond to an average value of four of more measurements, while the error bars indicate the high and the low values. Comparison of Figure 10 with Figure 6 reveals that the samples prepared by the hybrid photo/thermal method exhibit mechanical properties equivalent to the fully cured samples prepared using only a photoinitiator. More importantly, Figure 10 illustrates that the photo/thermal samples exhibit fully developed mechanical properties in approximately 3 minutes, compared to 7 minutes for the samples cured using only a photoinitiator. These results suggest that the hybrid photo/thermal initiation approach has tremendous potential for preparation of fiber-filled composites in relatively short cycle times.

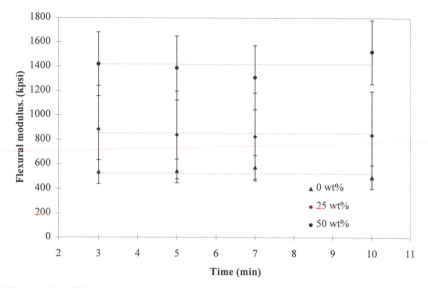

Figure 10. Flexural modulus as a function of illumination time for photocured composites with fiber loadings of 0, 25, and 50 wt.%.

Conclusions

We have presented a series of studies on photoinitiated polymerizations of vinyl ester glass fiber composites.   These studies demonstrate the feasibility of photopolymerizations of thick and fiber-filled polymers based upon self-eliminating light intensity gradient and hybrid photo/thermal cure strategies using common commercially-available initiators and resins.   Absorbance studies on the initiator revealed that BEE exhibits effective photobleaching in the near UV (300 - 390 nm) region of the spectrum resulting in self eliminating light intensity gradients.   The gradient elimination rate was found to increase with increasing light intensity but decrease with increasing initiator concentration.   Measurement of the *in situ* temperature profiles revealed the reaction temperature is high enough to ensure adequate polymerization and property development without leading to thermal degradation.   These results suggest that it is feasible to photopolymerize thick and fiber-filled polymers with no external heating or cooling.

Studies were performed to demonstrate the effects of process variables such as light intensity, cure time, initiator concentration, and fiber loading on the evolution of the mechanical properties of the polymers and composites.   Even with moderate incident light intensities (less than 500 mW/cm^2) and high fiber loadings (60 wt.% random fibers) the photopolymerizations proceed to completion in minutes and exhibit mechanical properties equivalent to samples prepared by traditional thermal means. Increasing the fiber loading not only lead to an increase in the flexural modulus, but also increased the time required for cure. An optimum initiator concentration for efficient property development was observed. A concentration of 0.2 wt.% maximized the flexural modulus obtained in five minutes of cure at an intensity of 480 mW/cm^2. Concentrations above this optimum lead to incomplete or nonuniform cure due to incomplete elimination of the light intensity gradient. Finally, increases in the incident light intensity lead to a greater flexural modulus at a given cure time indicating that, as expected, higher intensities lead to higher cure rates.

Studies were performed to investigate a dual cure strategy in which the heat liberated by a photo-induced polymerization leads to the dissociation of a thermal initiator, resulting in the production of additional active centers. The samples containing only the photoinitiator exhibited cure times that increase linearly with thickness (the time required for complete cure was controlled largely by the time required for elimination of the light intensity gradient), while the samples containing both the photoinitiator and the thermal initiator exhibit a much shorter cure time, nearly independent of thickness. At moderate light intensities, samples more than 1 cm. thick could be prepared in less than 3 minutes. These results suggest that the dual cure strategy is very promising for the production of thick polymers and composites since it offers both the temporal control of the start of the reaction afforded by the photopolymerization, as well as enhanced reaction rate and completeness of cure provided by the thermal initiation.

Literature Cited

1. Johnson, C.F. *Composites*; Engineered Materials Handbook, ASM International: Metals Park, OH, 1987; Vol. 1, pp 135-138.
2. McCarvill, W.T. *Composites*; Engineered Materials Handbook, ASM International: Metals Park, OH, 1987; Vol. 1, pp 564-568.
3. Coons L.S.; Rangarajan, B.; Godshall, D.; Scranton, A.B. In *Innovative Processing & Characterization of Composite Materials*; Gibson, R.T.; Chou, T.W.; Raju, P.K., Eds; ASME, New York,1995, NCA Vol. 20, pp 227-240.
4. Renault, T.; Ogale, A.A.; Dooley, R.L.; Bagchi, A.; Jara-Almonte, C.C. *SAMPE Quarterly*, **1991**, *22*, 19.
5. Ogale, A.A.; Renault, T.; Dooley, R.L.; Bagchi, A.; Jara-Almonte, C.C. *SAMPE Quarterly*, **1991**, *23*, 28.
6. Renault, T.; Ogale, A.A.; Drews, M.J. *Proc. ANTEC '93*, **1993**, 2352.

Chapter 16

Pigmented Coatings Cured with Visible Light

Barbara F. Howell[1,3], Anita de Raaff[2,4], and Thomas Marino[2]

[1]Carderock Division, Naval Surface Warfare Center, Annapolis Detachment,
Annapolis, MD 21402–5067
[2]Spectra Group, Ltd., 1722 Indian Wood Circle, Maumee, OH 43537

In order to develop an environmentally friendly, visible light curable, pigmented coating which can be applied for touch up purposes to minimally prepared steel surfaces, use of photobleaching photoinitiators has been investigated. Additional requirements are that the coating does not yellow during aging and that it can be applied to a cold surface. Photoinitiators containing BAPO [bis(2,6-dimethoxybenzoyl)-2,4,4-trimethylpentyl-phosphine oxide] and DIBF (5,7-diodo-3-butoxy-6-fluorone) were tested. DIBF can initiate curing by free radical reactions as well as by cationic means. It is used with an amine and the cationic photoinitiator OPPI (4-octyloxyphenylphenyl-iodonium hexafluoroantimonate) for radical initiation, but for cationic initiation only OPPI and DIBF are needed. A cycloaliphatic epoxy has been rapidly cured cationically with this pair and only a small amount (0.1%) of DIBF photoinitiator is required. Performance test results are excellent indicating that through cure has been achieved.

The purpose of work described here is to develop a coating which can be used on naval vessels for touch up purposes while a ship is underway. Radiation curable coatings are of special interest because of the extremely low level of volatile organic compounds emitted during curing. To be useful the coating should be Navy gray and suitable for application to cold surfaces. When used in confined spaces a further limitation is that the application should not produce ozone and for some Naval applications mercury is not

[3]Current address: 6694 Flamingo Road, Melbourne Village, FL 32904
[4]Current address: General Electric Corporation, Antwerpen, Belgium

allowed. Photobleaching photoinitiators (1), which are
activated by visible light, have recently become
commercially available. These substances photobleach during
the curing process which eliminates the yellowing of a
pigmented coating that would otherwise occur with a
photoinitiator absorbing light at long ultraviolet or short
visible wavelengths. Photobleaching also allows light to
penetrate more deeply into the coating. Photobleaching
photoinitiators investigated were BAPO [bis(2,6-
dimethoxybenzoyl)-2,4,4-trimethylpentylphosphine oxide,
mixed 1/4 with 2-hydroxy-2-methyl-1-phenyl-1-propanone
(HMPP)], as well as DIBF (5,7-diiodo-3-butoxy-6-fluorone).

Experimental[1]

Materials

   Oligomers and Resins. Oligomers and resins used were:
Ebecryl 3700 epoxy acrylate, Radcure Specialties, Inc.,
Smyrna, GA 30080;   Unicarb, which is 73% UVR 6128 (a
cycloaliphatic epoxy), 26.5% polyol and 0.5% Silwet L-7602
(See under miscellaneous compounds),   Union Carbide
Chemicals and Plastics Co., Inc., Danbury, CT 06817-0001;
and Heloxy 505 Modifier, Shell Chemical Co., Houston, TX
77210.

   Monomers. Monomers used were: IBOA (isobornyl
acrylate), TMPTA (trimethylol-propane triacrylate), and
TRPGDA (tripropylene glycol diacrylate) all three from
Radcure Specialties, Inc., Smyrna, GA 30080;   TMPTMA
(tripropylene glycol trimethacrylate, Sartomer Co. West
Chester, PA 19283;   Photomer 4017, 1,6-hexanediol
diacrylate, Henkel Corp., Ambler, PA 19002-3498;   and Epon
828, the diglycidyl ether of bisphenol A, Shell Chemical
Co., Houston, TX 77210 via a distributor.

   Photoinitiators. Photoinitiators used include: Darocur
1173 and Darocur 4265 which are 2-hydroxy-2-methyl-1-phenyl-
1-propanone;   Irgacure 907, which is 2-methyl-1-
[4(methylthio)phenyl]-2-(4-morpholinyl)-1-propanone(MMMP);
and CGI 1700 which contains BAPO [bis(2,6-dimethoxybenzoyl)-
2,4,4-trimethylpentylphosphine oxide] mixed 1/4 with HMPP (2-
hydroxy-2-methyl-1-phenyl-1-propanone) all from Ciba Geigy,
Hawthorne, NY 10532;   FX-512, a sulfonium photoinitiator,
3-M Industrial Chemical Products Div., St. Paul MN 55144-1000;
DIBF, 5,7-diiodo-3-butoxy-6-fluorone (H-Nu 470), Spectra
Group, Ltd., Maumee, OH 43537;   OPPI, 4-octyloxyphenyl-
phenyliodonium hexafluoroantimonate, G. E. Silicones,
Waterford, NY 12188;   and DIDMA, N,N-dimethyl-2,6-

[1]To describe procedures adequately, it is occasionally helpful to identify commercial
products and equipment. In no case does such identification imply Naval Surface
Warfare Center recommendation or endorsement, nor does it imply that the item is
necessarily the best available for the purpose.

diisopropylaniline, First Chemical Corp., Pascagoula, MS 39568-1427.

Photosensitizer. The photosensitizer ITX, isopropylthioxanthone, was obtained from Aceto Corp., Lake Success, NY 11042-1215.

Pigments. Pigments employed include TiONA TiO_2, SCM Chemicals, lot 25-JKSJ, Baltimore, MD 21202; lampblack, lot 97 Pfizer Minerals and Pigments Div., New York, NY 10017; manganese ferrite and copper chromite, Ferro Colors, Ferro Corp., Pittsburgh, PA 15204.

Miscellaneous compounds. Other materials used include: FC-171, fluorocarbon surfactant, 3M Industrial Chemical Products Division, St. Paul MN 55144-1000; Byk 306, Bykchemie USA, Wallingford, CT 06492; Polyol (poly-caprolactonetriol a polyester polyol), and Silwet L-7602 (polyalkylene oxide modified polydimethylsiloxane), both from Union Carbide Chemicals and Plastics Co., Inc., Danbury, CT 06817-0001.

Equipment. Lamps used include an RC-500 xenon lamp obtained from Xenon Corp., Woburn MA 01801; a Portacure 1000F/1500F mercury lamp, American Ultraviolet, Murray Hill, N.J. 07974; and a Fusion Systems F300 ultraviolet lamp system operated with a V-bulb.

Test Procedures. Coatings were tested by the following procedures. For crosshatch adhesion, ASTM-D3359, a special scribe was employed which had 11 sharp points spaced approximately 1 mm apart. The film was scored twice with this tool so that scratches were perpendicular. Special tape was then placed firmly over the scored area and pulled off. The number of squares of film coating remaining were counted, and this provided a numerical basis for measuring surface adhesion. No squares are removed for a satisfactory coating.
For impact tests, ASTM-02794, a ball of known mass was dropped a fixed height through a tube, impacting the coated metal panel. Indirect impact resistance is reported as the fraction of paint removable (pickoff) from the impact site when the panel is struck on the uncoated side. Direct impact resistance is reported as the pickoff at the impact site when the panel is struck on the coated side. Optimally, no coating is loosened by this test.
For the balanced beam-scrape adhesion test, ASTM-2197, weights were added to a commercial balanced beam tester. The kind of scratch produced by the stylus with the different loadings was recorded and served as a measure of film hardness. Coatings which are not scratched to bare metal by 500 g are desirable.
For the salt fog test, ASTM-B117, panels were placed on a rack in a salt fog spray for 500 or 1000 hours at a

temperature of 49 °C. Amount and kind of blistering was assessed by the method of ASTM D-714. The best coatings do not blister or rust during this exposure.

Blistering, ASTM-D-714, was assessed according to the number and size of blisters as illustrated in this ASTM procedure. Number 6 blisters are approximately 1 to 2 mm in diameter, and for dense blistering 50% or more of the panel surface is covered by blisters.

Test Panel Preparation. Grit-blasted, steel Q-panels, 3 x 5 in^2 were stored in protective paper inside a desiccator. Just prior to use they were rinsed with a 1/1 (by volume) mixture of mineral spirits and methyl ethyl ketone, allowed to dry, then rinsed with methyl ethyl ketone. Coatings were applied with a brush and thinned to uniform thickness with a number 24 helically wound (Meyer) rod.

Results and Discussion.

Early Investigations. Work has been conducted at the Carderock Division to develop a suitable coating for Naval applications such as touching up a rusty area inside a tank. For the coating to cure well it is necessary for some component of the mixture to absorb radiation at a wavelength longer than 380 nm below which titanium dioxide absorbs strongly, Figure 1. The first combination tried was ITX (isopropyl-thioxanthone) and MMMP (2-methyl-1-[4(methylthio)-phenyl]-2-(4-morpholinyl)-1-propanone). ITX has a maximum absorbance at 386 nm and is known to transfer energy efficiently to MMMP, a photoinitiator, which can trigger the free radical reaction of the acrylate containing resin components. This combination produced curing in the presence of pigment, but the coated panel became slightly yellowed after 500 h in the salt fog. Therefore use of other photoinitiators was investigated.

BAPO. The photoinitiator, BAPO, absorbs throughout the UV but also in the long UV and short visible light regions. It has the following structure,

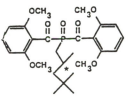

and its long UV-short Vis absorbance spectrum in shown in Figure 2. When exposed to light, it dissociates at the carbonyl groups forming two free radicals and one biradical. After dissociation, it loses its yellow color and initiates a chain reaction with the acrylate group which links molecules together producing the cured coating.

In addition to proper reagent selection, in order to

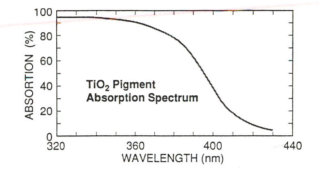

Figure 1. Titanium dioxide absorption spectrum.

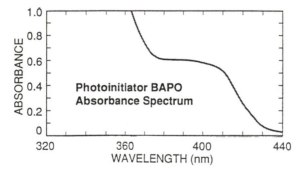

Figure 2. BAPO absorbance spectrum.

produce a pigmented coating which can be readily cured with ultraviolet or visible light it is necessary to choose a lamp which emits strongly at a wavelength that is not absorbed by the titanium dioxide pigment, that is at wavelengths longer than approximately 380 nm. A xenon lamp has a strong output in the ultraviolet above this wavelength and emission remains strong throughout the visible region. However, commercially available xenon lamps are more expensive than other kinds of lamps, and a reasonably priced unit cures a patch 2 x 3 in^2 which makes it impractical for cure of a much larger area such as 2 x 2 ft^2.

A mercury lamp is commercially available which can cure an area 4 x 10 in^2, and it has a lamp unit that is light weight enough to be hand held so that it can be moved over a surface to cure a coating. Moreover, ozone free lamps are available for this unit which eliminates problems with ozone generated when work is done in a confined space. However, mercury lamps have a weak output in the long UV-short Vis spectral region, Figure 3, and for some Naval purposes, mercury is not allowed.

Another possible lamp alternative is a Fusion Systems lamp for which a variety of bulbs are available with spectral outputs in different spectral regions. Spectra of the bulbs available for Fusion Systems lamps are shown in Figure 3. By comparing Figures 1, 2, & 3, it can be seen that the V-bulb emits in a spectral region in which titanium dioxide does not absorb strongly and that the BAPO absorbs radiation at these wavelengths so that light can penetrate into the coating. A gray-pigmented epoxy acrylate formulation has been successfully cured with use of BAPO/HMPP and the V-bulb.

Initially, carbon black was added to the UV-curing formulation to produce a gray color, but it was found that this substance is a free radical absorber which interferes with the UV-curing process. Following a suggestion made by Lucie (Microlite Corp., personal communication, 1990.) manganese ferrite black spinel was used instead of carbon black to produce the gray color when mixed with titanium dioxide. It was found that when spinel comprised 10% of the pigment, the desired gray color was produced, and curing occurred much more readily than with the carbon black. Dispersion of the pigment was done by ballmilling into a mixture of the epoxy acrylate (Ebecryl 3700) thinned 2/1 by weight with reactive diluent.

Since the commercially available epoxy acrylate is extremely viscous, it is necessary to thin the formulation with a reactive diluent. A mixture with satisfactory viscosity for brush-on application contains about 45% diluent. Of the diluents tested, TMPTA (trimethylolpropane triacrylate) proved most successful.

The formulation which performed best contained 1% BAPO/HMPP photoinitiator, 50% acrylate ester of bisphenol A epoxy (Ebecryl 3700), 45% reactive diluent, 5% pigment, and a few drops of the fluorocarbon surfactant FC-171.

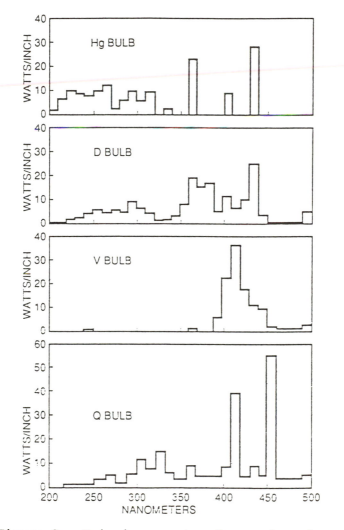

Figure 3. Emission spectra for various lamps

Several reactive diluents were tested which include trimethylolpropane triacrylate (TMPTA), tripropylene glycol diacrylate (TRPGDA), isobornylacrylate (IBOA), and 1,6-hexanedioldiacrylate (HDODA). Since TMPTA is a triacrylate, it provides more crosslinking than the other diluents and produces a very hard glossy coating. However with TMPTA, it was difficult to get good through cure, especially if the coating thickness exceeded 60 micrometers. Through cure was measured with the cross hatch adhesion test. If the coating has not been cured next to the metal surface it will be removed by the tape is applied for this test. Control of coating thickness, as is done with the Meyer rod, is essential. The Meyer rod has a solid metal core which is helically wrapped with wire of known diameter. A number 24 rod gives a cured coating about 50 micrometers thick. Through cured TMPTA containing formulations performed well in impact, balanced beam, crosshatch adhesion, and salt fog tests as shown in Table I. It also successfully withstood six months exposure to salt water or to UV light without loss of coating integrity, although chalking was observed. These results are satisfactory for the intended application.

It was found, when an attempt was made to conduct ship trials that the paint applier would be required to wear long sleeves, gloves, safety glasses, and a protective face shield. It is also a requirement that an eyewash fountain be nearby. These requirements were imposed because acrylates are skin, eye and respiratory tract irritants, but the problem lies mostly with the acrylated reactive diluent rather than with the acrylated resin (2,3). To avoid these problems, tests have been performed with use of trimethylolpropane trimethacrylate (TMPTMA) as a replacement for TMPTA. TMPTMA has been shown to be much less of an irritant than TMPTA, and is used in dental restorations.

When used as part of the radiation curing coating mixture, TMPTMA was found to cure more slowly than TMPTA and it is more difficult to get through cure. After testing many photoinitiator combinations, a successful formulation was produced which contained ITX, BAPO/HMPP, and N,N-dimethyl p-toluidine combined one part to 1.7 parts hexanediol diacrylate by weight. This combination made up 6.5% of the total UV curing mixture which also contained 40 percent TMPTMA, 48.5% epoxy acrylate and 5% pigment by weight.

Although the V-bulb works well with BAPO/HMPP, it is a metal halide bulb and contains mercury. In addition it is microwave powered, which provides quick start up and high light intensity, but a high speed fan is required for adequate cooling. This makes the lamp assembly cumbersome and heavy (it weighs sixteen pounds). In addition shielding from microwave radiation is provided by covering the end of the lamp assembly unit with a very fine tungsten wire mesh. The mesh could be easily broken if the lamp unit were to be moved about to cure a vertical surface so that the protection from microwaves would be lost.

Table I. Comparison of Resins and Photoinitiators

	Crosshatch Adhesion (Percent removed)[a]	Impact test[b] Front Back (in-lbs)	Balanced Beam[c] (Mass to scratch to bare metal)	Salt Fog[d] 500 h, 49 °C
Epoxy acrylate BAPO/HMPP TMPTA	0	40 30 % pickoff 0 50	200g,	No. 2 medium dense Least amount of rust
Epoxy acrylate, DIBF[e,f] TMPTA	5%	40 30 % pickoff 10 10	500g	Few to medium no. 2 blisters A little rust around the edges Bleaching
Cycloaliphatic epoxy III DIBF[g]	0%	60 60 % pickoff 5 50	400g	Overall wrinkling Moderate rusting Dense #6 blisters Bleaching

[a]ASTM-D3359, [b]ASTM-D2794, [c]ASTM-2197, [d]Salt fog--ASTM-B117;Blistering--ASTM-D714.
[e]DIBF is 5,7-diiodo-3-butoxy-6-fluorone. [f]Cured with a V-bulb. [g]Cured with a tungsten halogen bulb.

DIBF. Spectra Group, working in conjunction with CDNSWC, has conducted many tests with the photoinitiator DIBF originated at the Bowling Green State University by Neckers and coworkers (4). DIBF (5,7-diiodo-3-butoxy-6-fluorone, H-Nu 470) is of special interest because the very large molar absorption coefficient at 470 nm 30,200 allows low concentrations to initiate photopolymerization (4). It has the following chemical structure.

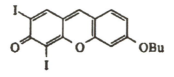

DIBF has the absorbance spectrum shown in Figure 4. Coating mixtures containing this photoinitiator can be cured with an intense tungsten halogen lamp, such as a projector bulb. When DIBF is irradiated it forms an excited state which has a very low absorbance, i.e., it photobleaches (1). This allows light to penetrate deeply into the coating.

Because visible light is not energetic enough to break chemical bonds, direct production of free radicals by the photoinitiator does not occur. Instead when cationic initiation is needed, as for reaction with epoxies, DIBF is used in conjunction with an iodonium compound such as 4-octyloxyphenyl-phenyliodonium hexafluoroantimonate (OPPI). It has been proposed that when irradiated, DIBF and OPPI interact to form a cationic species.

The DIBF OPPI combination has been shown to efficiently cure a wide variety of epoxies including cycloaliphatics. With this photoinitiator it is possible to cure bisphenol A epoxies such as Epon 828 quickly without the need for acrylation of the epoxy. Cyclo-aliphatic epoxies were of special interest because they were expected to react much faster than bisphenol A type epoxies. Those tested include 3,4-epoxycyclohexylmethyl-3',4'-epoxycyclohexyl-carboxylate (UVR 6110), bis(3,4 epoxy-cyclohexylmethyl) adipate (UVR 6128), and 1,2-epoxy-4-vinylcyclohexane (vinyl cyclohexene oxide). It was found that the vinyl cyclohexene oxide reacted rapidly, but work with it was discontinued because it has a fairly high vapor pressure (2 torr at 20 $^{\circ}$C), an intense odor, and the photoinitiator does not dissolve in this resin.

A promising formulation was produced which contains the cycloaliphatic epoxy UVR 6110. The formulation contains 73% of the epoxy, 26.5% Polyol, 5% Silwet L7602 0.1% DIBF, 2.5% OPPI and 5-10 drops FC-171 in 100 g of

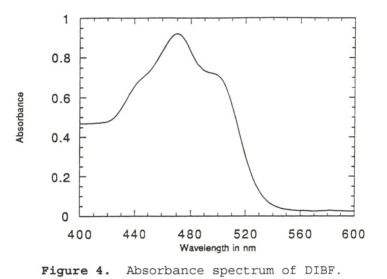

Figure 4. Absorbance spectrum of DIBF.

resin with 5% pigment. Coating layers 100 micrometers
thick could be cured in open air with the V-bulb and 4
passes under the lamp when the belt speed was 6 ft/min.
        The UVR 6110 formulation was applied to grit blasted
steel Q panels. After cure it was found that with the
crosshatch adhesion test there was no pulloff. In the
impact test (D2794), with 6.8 j of impact, 95% of the
coating remained on the front impact and 50% on the back
impact surface. With the balanced beam (ASTM D2197), 400
g were required to scratch to bare metal. These results
are considered satisfactory. However, in salt fog (49 $^{\circ}$C,
1000 h) the coating performed less well. At the end of
1000 h, the panels showed overall wrinkling, moderate
rusting, dense #6 blisters and bleaching. By way of
comparison, a standard two part epoxy coating shows no
wrinkling, rusting or blistering under these conditions.
        Another formulation tested contained 30% of the UVR
6110 formulation (above) and 70% of a mixture consisting of
73% UVR 6128, 26.5% Polyol and 0.5% Silwet L7602 with the
same percentages of the FC-171 and pigment as the first
formulation. There was no removal of coating in cross
adhesion, no loosened coating after 6.8j impact, and 1000g
were required on the balanced beam to scratch to the metal
surface, satisfactory results. However, after 500 h in the
salt fog there were many rust streaks and #6 dense or
medium dense blisters; unsatisfactory performance. In
addition there was bleaching of the pigment.
        Bisphenol A type epoxies such as Epon 828, were also
cured with the DIBF OPPI combination. When a modifier,
Heloxy 505 (a low viscosity polyepoxide modifier) was
added, viscosity was reduced and adhesion to the metal
surface and impact resistance were improved as compared
with the bisphenol A only. Surprisingly cure was faster
than with the Union Carbide cycloaliphatic resins, but
charring of the resin during cure was a problem.
        When used with N,N-dimethyl-2,6-diisopropylaniline
(DIDMA), the DIBF-OPPI photoinitiator can be used to cure
systems which react by a free radical mechanism. It has
been suggested that the amine donates an electron to the
DIBF converting it to the radical anion while forming an
amine radical cation which then loses H^{+} to become an amine
free radical. Although the OPPI is known to enhance
radical curing, its role has not been clearly defined. The
amine radical species formed upon irradiation reacts with
acrylate double bonds forming acrylate radicals which cause
cross linking (curing) (4).
        Epoxy acrylate (Ebecryl 3700, 50%) combined with 25%
TMPTA, 25% IBOA and with DIBF-OPPI-DIDMA pigment and FC-171
as above was also used to coat panels. These panels showed
no removal of coating in the cross hatch adhesion test.
With 6.8j there was no loosened coating on the frontal
impact surface, and only 30% was loosened on the back
surface. Five hundred grams were required to scratch to
bare metal with the balanced beam. Performance was

satisfactory for these three tests as was found for the cycloaliphatic epoxies. However, after 1000 h in the salt fog, there was extensive rust bleed and dense #6 blistering as well as bleaching of the pigment, unsatisfactory performance.

A problem with this photoinitiator is that curing is very slow with commercially available lamps such as a mercury lamp or a V-bulb, bulbs which do not emit strongly at the 470 nm absorbance maximum of DIBF. A Q bulb works best, but is not optimal. At present there is no commercially available, intense visible lamp for curing areas as large as 4 x 6 in^2 quickly. Therefore, work is being undertaken to develop a tungsten halogen lamp which will cure a larger area. A sulfur lamp is being considered as an alternative as well. The curing lamp should be portable and should not contain mercury. This type of lamp will not pose the eye hazard associated with a UV emitting bulb.

Borate photoinitiators.

Workers at Spectra Group, Ltd. (5) have recently investigated use of borate compounds (tetramethylammonium triphenylbutyl borate (1), and butyryl choline triphenyl butylborate (5) in conjunction with DIBF, OPPI and DIDMA.

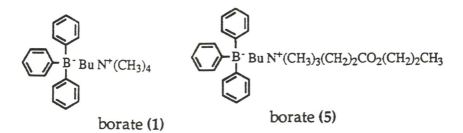

borate (1)   borate (5)

These borate compounds act as electron donors and interact with DIBF generating a butyl radical which can initiate acrylate polymerization. With a urethane acrylate, eighty seven percent acrylate double bond conversion has been found to occur with 5 sec exposure to a 75 W tungsten bulb when weight percents of the three photoinitiator components are 0.05% DIBF, 0.43% OPPI and 0.5% borate 5. Comparable results were obtained with epoxy acrylates. Borate 1 is less soluble in an acrylate formulation than is borate 5 which may account for the greater effectiveness found for borate 5 in increasing curing rate. Although no coating performance tests have been performed following borate assisted curing, the performance is expected to be comparable to that of other acrylate coatings.

Conclusions.

Photobleaching visible light photoinitiators have been tested for cure of gray pigmented coatings and found to produce coatings which meet performance requirements. The best results so far have been obtained by use of BAPO/HMPP in an epoxy acrylate. Although work with DIBF has not been optimized, it is clear that it is feasible to cure radiation curable coatings with visible light such as that emitted by a tungsten halogen lamp.

Use of visible light is desirable if a portable light curing source is to be used because of the reduced eye hazard and the fact that these lamps do not contain mercury. However, at the present time there is no commercially available portable light source meeting the performance requirements.

Since acrylates are skin, eye and respiratory system irritants, it would be desirable to find a system which did not contain these groups. Methacrylates are being tested as possible alternatives, but cure is much slower than with acrylates.

Acknowledgement.

Funding support for this work has been received from the Office of Naval Research

Literature Cited.

1. Pappas, S. P., In *Radiation Curing Science and Technology*; Pappas, S. P., Ed., Plenum Press: New York, NY, 1992, p 9.
2. *American Industrial Hygiene Assoc. J.* **1981**,*42*, pp853 ff.
3. Andrews, L. S.; Clary, J. J. *J. Toxicology & Env. Health* **1986**,
4. Marino, T. L.; Martin, D.; Neckers, D. C. *Adhesives Age*, *1995*, *pp22 ff.*
5. Marino, T. L.; De Raaff, A. M.; and Neckers, D. C. *RadTech'96 North America UV/EB Conference Proceedings*, April 28 - May 2, 1996, pp 7-16.

INDEXES

Author Index

Affiliation Index

Subject Index

Bestsellers from ACS Books

The ACS Style Guide: A Manual for Authors and Editors (2nd Edition)
Edited by Janet S. Dodd
470 pp; clothbound ISBN 0–8412–3461–2; paperback ISBN 0–8412–3462–0

Writing the Laboratory Notebook
By Howard M. Kanare
145 pp; clothbound ISBN 0–8412–0906–5; paperback ISBN 0–8412–0933–2

Career Transitions for Chemists
By Dorothy P. Rodmann, Donald D. Bly, Frederick H. Owens, and Anne-Claire Anderson
240 pp; clothbound ISBN 0–8412–3052–8; paperback ISBN 0–8412–3038–2

Chemical Activities (student and teacher editions)
By Christie L. Borgford and Lee R. Summerlin
330 pp; spiralbound ISBN 0–8412–1417–4; teacher edition, ISBN 0–8412–1416–6

Chemical Demonstrations: A Sourcebook for Teachers, Volumes 1 and 2, Second Edition
Volume 1 by Lee R. Summerlin and James L. Ealy, Jr.
198 pp; spiralbound ISBN 0–8412–1481–6
Volume 2 by Lee R. Summerlin, Christie L. Borgford, and Julie B. Ealy
234 pp; spiralbound ISBN 0–8412–1535–9

From Caveman to Chemist
By Hugh W. Salzberg
300 pp; clothbound ISBN 0–8412–1786–6; paperback ISBN 0–8412–1787–4

The Internet: A Guide for Chemists
Edited by Steven M. Bachrach
360 pp; clothbound ISBN 0–8412–3223–7; paperback ISBN 0–8412–3224–5

Laboratory Waste Management: A Guidebook
ACS Task Force on Laboratory Waste Management
250 pp; clothbound ISBN 0–8412–2735–7; paperback ISBN 0–8412–2849–3

Reagent Chemicals, Eighth Edition
700 pp; clothbound ISBN 0–8412–2502–8

Good Laboratory Practice Standards: Applications for Field and Laboratory Studies
Edited by Willa Y. Garner, Maureen S. Barge, and James P. Ussary
571 pp; clothbound ISBN 0–8412–2192–8

For further information contact:

American Chemical Society
1155 Sixteenth Street, NW ◆ Washington, DC 20036
Telephone 800–227–9919 ◆ 202–776–8100 (outside U.S.)

The ACS Publications Catalog is available on the Internet at
http://pubs.acs.org/books

Highlights from ACS Books

Desk Reference of Functional Polymers: Syntheses and Applications
Reza Arshady, Editor
832 pages, clothbound, ISBN 0–8412–3469–8

Chemical Engineering for Chemists
Richard G. Griskey
352 pages, clothbound, ISBN 0–8412–2215–0

Controlled Drug Delivery: Challenges and Strategies
Kinam Park, Editor
720 pages, clothbound, ISBN 0–8412–3470–1

Chemistry Today and Tomorrow: The Central, Useful, and Creative Science
Ronald Breslow
144 pages, paperbound, ISBN 0–8412–3460–4

Eilhard Mitscherlich: Prince of Prussian Chemistry
Hans-Werner Schutt
Co-published with the Chemical Heritage Foundation
256 pages, clothbound, ISBN 0–8412–3345–4

Chiral Separations: Applications and Technology
Satinder Ahuja, Editor
368 pages, clothbound, ISBN 0–8412–3407–8

Molecular Diversity and Combinatorial Chemistry: Libraries and Drug Discovery
Irwin M. Chaiken and Kim D. Janda, Editors
336 pages, clothbound, ISBN 0–8412–3450–7

A Lifetime of Synergy with Theory and Experiment
Andrew Streitwieser, Jr.
320 pages, clothbound, ISBN 0–8412–1836–6

Chemical Research Faculties, An International Directory
1,300 pages, clothbound, ISBN 0–8412–3301–2

For further information contact:

American Chemical Society
Customer Service and Sales
1155 Sixteenth Street, NW
Washington, DC 20036

Telephone 800–227–9919
202–776–8100 (outside U.S.)

The ACS Publications Catalog is available on the Internet at
http://pubs.acs.org/books